High-Quality
Industrial Water
Management Manual

High-Quality Industrial Water Management Manual

Paul N. Garay
and
Franklin M. Cohn

The Fairmont Press
700 Indian Trail
Lilburn, Georgia 30247

Library of Congress Cataloging-in-Publication Data

Garay, Paul N., 1913-
   High-quality industrial water management manual / Paul N. Garay
and Franklin M. Cohn.
   p. cm.
   Includes bibliographical references and index.
   ISBN 0-88173-087-4
   1. Water quality management. 2. Water-supply, Industrial.
I. Cohn, Franklin M., 1923- . II. Title.
TD365.G37   1992   628.1-dc20   89-45174
                                        CIP

High-Quality Industrial Water Management Manual.

Published by The Fairmont Press, Inc.
700 Indian Trail
Lilburn, GA 30247

Printed in the United States of America

10 9 8 7 6 5 4 3 2 1

ISBN 0-88173-087-4   FP

ISBN 0-13-389560-2   PH

While every effort is made to provide dependable information, the publisher, authors, and editors cannot be held responsible for any errors or omissions.

Distributed by Prentice-Hall, Inc.
A Simon & Schuster Company
Englewood Cliffs, NJ 07632

Prentice-Hall International (UK) Limited, London
Prentice-Hall of Australia Pty. Limited, Sydney
Prentice-Hall Canada Inc., Toronto
Prentice-Hall Hispanoamericana, S.A., Mexico
Prentice-Hall of India Private Limited, New Delhi
Prentice-Hall of Japan, Inc., Tokyo
Simon & Schuster Asia Pte. Ltd., Singapore
Editora Prentice-Hall do Brasil, Ltda., Rio de Janeiro

Acknowledgments

In the preparation of this manual, information and data were extracted and derived from available relevant technical literature on the subject. Credit is given to the following sources.

Corporate Publications

Betz Laboratories
Calgon Corporation
Dearborn Chemical Company
Diamond Shamrock Chemical Company
Dow Chemical Company
Drew Industrial Division of Ashland Oil Corp.
Nalco Chemical Company
Rohm & Haas
Sybron Corporation
Wright Chemical Corporation

Technical Journals

Combustion
Industrial Water Engineering

Contents

Foreword

This is not a text on water treatment. There are no guides contained describing specifically how to modify water. This is, however, a source-book that describes the kinds of problems encountered with water, where to look for answers, how to recognize the potential of water for specific uses. Above all, it is a manual to provide information to executives and operators regarding the management of water resources.

Water, a compound of hydrogen and oxygen, is, in its pure state, a tasteless, odorless, colorless liquid. In its normal, less than pure state, it constitutes our rivers, lakes, oceans, ground aquifers, and rain. It is our most universal compound without which there would be no life as we know it, on earth, and very little chemical activity. In this text, we are interested in biological and industrial uses. While the role of water in industry is less important than its role in supporting life, where it is indispensable, it is both ironic and fortunate that industry requires the chemically purer water.

With the expansions of urban centers in populated areas, and the proliferation of industry in the same areas, it is becoming more and more difficult to find water of the quality and quantity needed to adequately supply the combined biologic and industrial requirements. Impurities in the water can result in ill health in humans and animals, and in corrosion, fouling and failure of equipment in industry. The degree to which natural water must be treated to become suitable for its varied uses, is the subject of past and present concern.

Management of water resources is a very broad subject that cannot be treated comprehensively in any one volume such as this text. This manual, however, is intended to provide plant managers and engineers with an appreciation of water quality related problems, and with some techniques for providing acceptable solutions to these problems.

Virtually every industry, in some way, requires or benefits from properly treated water for use as a solvent, cleaning medium, heat exchange agent, thermodynamic fluid, ingredient, or catalyst.

It would be difficult to find a product or a process, to which tainted water would not be detrimental. Rayon, cotton, silk, wool, glass, paper, leather, hard and soft beverages, ice, pharmaceuticals, electronics, ceramics, plastics, and innumerable other products, all require water in their manufacture. And in processes, such as metal-plating, laundering, food preparation, mirror silvering, television tube production, and photography, water quality is equally important—just as it is in steam plants, engines, and power plants.

For chemically acceptable water, all of these depend on water treatment. Water treatment to remove or alter the character of impurities is an essential process without which the industrial progress of the past century would have been impossible.

This text contains a discussion of water-related problems and their mitigation with respect to all these uses. The reader may note some duplication of information. This is useful to present the material from more than one viewpoint.

There are also some older articles included. This has been done because the information is so well presented that the reader may gain insight into the way in which different problems are related.

1

Overview

Water, in the various aspects of the hydrologic cycle, is indeed the basis for all life on the planet. The primordial soup in which life first originated was largely water. Today, animal and vegetable life are sustained by this essential fluid. Next to air, water is the substance with which we are most intimately acquainted. It is a solvent without equal. Because it is so familiar, we readily overlook the fact that it is a unique substance. Behaving unlike any other liquid, it has the rare property of being denser as a liquid than as a solid, being most dense at 4 degrees Celsius; this property makes it possible for fish to exist at the lower levels of a frozen pond, since water at the bottom normally does not freeze.

The elements of which water is composed—oxygen and hydrogen— are chemically exceptional. Both are unusually reactive. Pure water, chemically labelled as H_2O, is the product of the combination of two hydrogen atoms (H_2) and one oxygen atom (O). The H_2O molecules are joined together so that, in effect, the whole mass of water in a vessel can be considered to act as a single molecule. Because of this bonding, one might expect the viscosity of water to be unusually high. Not so, however, since each hydrogen bond is shared, usually between two other molecules, and one of these bonds is easily broken.

Water rarely exists in a pure form. All natural waters contain varying amounts of dissolved and suspended matter. The type and

1

amount of matter in water varies with the source. The pure form probably is represented by the moisture in clouds. When moisture is precipitated as rain, however, on the way down it collects atmospheric dust, carbon dioxide, and atmospheric gases, chiefly oxygen and nitrogen. Water on the surface picks up organic material from plant life, as well as various microorganisms. Percolating through the soil, it dissolves and picks up in suspension and solution many minerals. Even ground water may absorb volatile organic material. Consider that eons ago, when the oceans were new, all water was fresh. In the ensuing billions of years, minerals leached out of the soil and were washed down into the oceans. Common ocean water contains about 32,000 parts per million (ppm) of minerals, the largest component being sodium chloride (common salt). A lesser content of dissolved salt, say about 7,000 ppm, is termed "brackish."

Solid material dissolved in water does not always retain its molecular status. It may break up into its constituent atoms, which are then known as "ions." Each ion, characterized by an electrical charge, called "valence," is either positive (+) or negative (−). Negative charges are called "cations," while positive ions are termed "anions." When removed from water, the various ions unite in combinations which result in a net electrical charge of zero. For instance:

Cations	Combination	Anions
Hydrogen, H, +1	H_2O	Hydroxide, OH, −1
Sodium, Na, +1	NaCl	Chloride, Cl, −1

There are many other examples of chemical combinations. The exact numbers of participating ions can be worked out on the basis of experience and analysis. It should be noted that all ions are not single atoms. Frequently combinations develop, such as the hydroxide ion, which is more commonly referred to as a "radical." Other radicals may be HCO_3, the bicarbonate radical, CO_3, and a host of others.

Suspended solids are those which are not dissolved in water and can be removed by settlement or filtration. Examples are mud, silt, clay, and metallic oxides. Dissolved solids are those in solution which cannot be removed by ordinary filtration or settlement. Major dissolved materials in water are silica, iron, calcium, magnesium, potassium, and sodium, in various compounds. Metallic constituents occur as bicarbonate, carbonate, sulfate, and chloride radicals.

Practically all commercial, industrial, and sanitary processes require water. The grade or purity may vary, depending on the ultimate use, but almost all commercial and industrial uses require some degree of treatment for optimum process use. High-purity water is required for activities such as steam generation and manufacturing processes. Relatively high concentrations of impurities can be tolerated for many uses, if the user is willing to accept the concomitant problems. For example, relatively high concentrations may be found in open cooling water recirculating systems. High levels of impurities, however, in cooling water systems can result in serious operating problems: deposit formation in heat exchangers, corrosion of metals, biological fouling, and wood deterioration in cooling towers. To minimize such problems significant costs may be incurred for internal chemical treatment.

Water treatment methods, therefore, require attention since:

- contamination cannot be tolerated in many processes;
- energy conservation benefits can result from appropriate treatment;
- the frequency of operational problems and the need for repairs caused by "bad" water can be significantly reduced;
- environmental pollution that may result from water treatment system discharges must be addressed;
- in nuclear uses, radiation may occur from irradiated solids in the water; and
- water treatment requirements and treatment programs for steam generators, air compressors, engines, condensers, cooling towers, and refrigeration systems are usually poorly defined.

In general, the processes that subject water to the greatest temperatures and pressures usually require the highest quality of water. Note the specific uses, in decreasing order of required purity are:

- electronic manufacturing processes, medical and nuclear uses;
- high pressure steam plants;
- industrial process systems;
- medium and low pressure steam systems;
- human uses; and
- open recirculating cooling water and cooling tower systems.

The direct results of poor quality water are corrosion, fouling, and scale deposits. All water-related problems stem from these factors.

Corrosion is an electrochemical process that dissolves metal. Deterioration and deposits caused by corrosion can result in failure of essential components such as boiler tubes, steam lines, air compressor coolers, and cooling water systems, to name but a few of the systems affected. Problems that result from inadequate water treatment, or improper treatment, are the cause of significant costs in avoidable repair work and equipment downtime.

Scale deposits are solids and organic growths that adhere to all surfaces wetted by the water, such as boiler tubes. In a heat transfer application, as in a heat exchanger or a boiler tube, scale acts as an insulator and reduces the rate of heat transfer. Refer to Figure 1-1. Inhibiting heat transfer causes two problems: loss of efficiency, with resulting increase in fuel consumption, and possible elevation of tube temperatures to the point where the metal weakens, and the tube ruptures. Scale build-up also increases pressure drop, and decreases water flow in a pipe. This clogging effect results in an increase of pumping power required to circulate the water, again with loss of efficiency.

Figures 1-2 through 1-5 illustrate corrosion and scale problems in steam plants and cooling tower components. Table 1-1 displays difficulties caused by water impurities, and means of treatment. Table 1-2 illustrates general problems caused by impurities in water with regard to specific uses. Table 1-3 generically illustrates the treatment of water from rivers, ponds, and wells.

Various degrees of water quality are required by specific uses. Figure 1-6 indicates the grade required for various uses, based on conductivity. Figure 1-7 illustrates the molecular/atomic spectrum of water impurities, and indicates the type of equipment used for removal of dissolved and particulate solids.

The impurities which must be removed from water depend, naturally, upon the water source, and the intended uses. All raw water, whether drawn from wells, rivers, lakes or other sources, contains chemical and bacterialogical contaminants, dissolved and suspended solids washed from the land, from sewer outlets, factory drainage, or dissolution of plant material in marshes and swamps.

Inland waters differ from those in coastal regions; well water is different from river water, and river water from lake water. Two plants along the same body of water, perhaps separated by a few miles, may have water supplies of very dissimilar characteristics. Even the concentration and character of dissolved material may vary with changing seasons and weather in the same locality. Often raw water must be filtered, softened, as a minimum treatment, and perhaps de-alkalized. In certain areas, sulfides and fluorides must be removed. Iron products are very common. In recent years, in many industrial locations, ground water has become contaminated with volatile organic compounds (VOC). For the most sophisticated uses, all ions must be removed by deionization, reverse osmosis, or some combination of treatments.

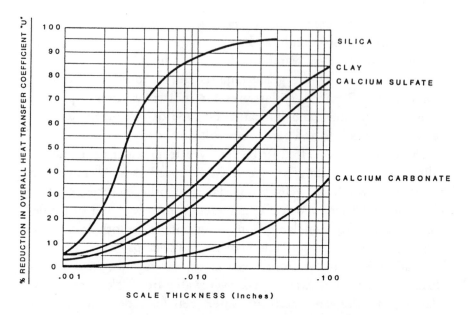

Figure 1-1. Influence of Scale on Heat Transfer

Figure 1-2. Section of Condensate Line Destroyed by Carbon Dioxide (Low ph) Corrosion. Metal destruction, resulting in thinning, is spread over a relatively wide area.

Figure 1-3. Scale Build-Up in Cooling Water Lines (High pH)

Figure 1-4. Calcium Phosphate Scale on Cooling Tower Fill Members

Figure 1-5. Tube Blister Due to Periodic Overheating
Note fissure and creep cracks, typical result of alternating
overheating and cooling.

Table 1-1. Difficulties Caused by Water Impurities

Constituent	Chemical Formula	Difficulties Caused	Means of Treatment
TURBIDITY	None—expressed in analysis as units.	Imparts unsightly appearance to water. Deposits in water lines, process equipment, etc. Interferes with most process uses.	Coagulation, settling and filtration.
COLOR	None—expressed in analysis as units.	May cause foaming in boilers. Hinders precipitation methods such as iron removal and softening. Can stain product in process use.	Coagulation and filtration. Chlorination. Adsorption by activated carbon.
HARDNESS	Calcium and magnesium salts expressed as $CaCO_3$.	Chief source of scale in heat exchange equipment, boilers, pipe lines, etc. Forms curds with soap. Interferes with dyeing, etc.	Softening. Demineralization. Internal boiler water treatment. Surface-active agents.
ALKALINITY	Bicarbonate (HCO_3), carbonate (CO_3), and hydrate (OH), expressed as $CaCO_3$.	Foaming and carryover of solids with steam. Embrittlement of boiler steel. Bicarbonate and carbonate produce CO_2 in steam, a source of corrosion in condensate lines.	Lime and lime-soda softening. Acid treatment. Hydrogen zeolite softening. Demineralization. Dealkalization by anion exchange.
FREE MINERAL ACID	H_2SO_4, HCl, etc. expressed as $CaCO_3$.	Corrosion.	Neutralization with alkalies.
CARBON DIOXIDE	CO_2	Corrosion in water lines and particularly steam and condensate liens.	Aeration. Deaeration. Neutralization with alkalies.
pH	Hydrogen ion concentration defined as $pH = Log\dfrac{1}{(H^+)}$	pH varies according to acidic or alkaline solids in water. Most natural waters have a pH of 6.0-8.0.	pH can be increased by alkalies and decreased by acids.

Table 1-1. Difficulties Caused by Water Impurities (concluded)

Constituent	Chemical Formula	Difficulties Caused	Means of Treatment
SULFATE	$(SO_4)^{--}$	Adds to solids content of water, but in itself, is not usually significant. Combines with calcium to form calcium sulfate scale.	Demineralization.
CHLORIDE	Cl^-	Adds to solids content and increases corrosive character of water.	Demineralization.
NITRATE	$(NO_3)^-$	Adds to solids content, but is not usually significant industrially. High concentrations cause methemoglobinema in infants. Useful for control of boiler metal embrittlement.	Demineralization.
FLUORIDE	F^-	Cause of mottled enamel in teeth. Also used for control of dental decay. Not usually significant industrially.	Adsorption with magnesium hydroxide, calcium phosphate, or bone black. Alum coagulation.
SILICA	SiO_2	Scale in boilers and cooling water systems. Insoluble turbine blade deposits due to silica vaporization.	Hot process removal with magnesium salts. Adsorption by highly basic anion exchange resins, in conjunction with demineralization.
IRON	Fe^{--} (ferrous) Fe^{---} (ferric)	Discolors water on precipitation. Source of deposits in water lines, boilers, etc. Interferes with dyeing, tanning, papermaking, etc.	Aeration. Coagulation and filtration. Lime softening. Cation exchange. Contact filtration. Surface-active agents for iron retention.

MANGANESE	Mn^{--}	Same as iron.	Same as iron.
OXYGEN	O_2	Corrosion of water lines, heat exchange equipment, boilers, return lines, etc.	Deaeration. Sodium sulfite. Corrosion inhibitors.
HYDROGEN SULFIDE	H_2S	Cause of "rotten egg" odor. Corrosion.	Aeration. Chlorination. Highly basic anion exchange.
AMMONIA	NH_3	Corrosion of copper and zinc alloys by formation of complex soluble ion.	Cation exchange with hydrogen zeolite. Chlorination Deaeration.
DISSOLVED SOLIDS	None	"Dissolved Solids" is a measure of total amount of dissolved matter, determined by evaporation. High concentrations of dissolved solids are objectionable because of process interference and as a cause of foaming in boilers.	Various softening process, such as lime softening and cation exchange by hydrogen zeolite will reduce dissolved solids. Demineralization.
SUSPENDED SOLIDS	None	"Suspended Solids" is the measure of undissolved matter, determined gravimetrically. Suspended solids cause deposits in heat exchange equipment, boilers, water lines, etc.	Subsidence. Filtration, usually preceded by coagulation and settling.
TOTAL SOLIDS	None	"Total Solids" is the sum of dissolved and suspended solids, determined gravimetrically.	See "Dissolved Solids" and "Suspended Solids."

Source: Betz Laboratories

Table 1-2. General Problems Caused by Impurities in Water

	A	B	C	D	E	F	G	H
Uniform corrosion	×	×	×	×	×	×	×	×
Localized corrosion	×	×	×	×	×	×	×	×
Stress corrosion	—	—	—	—	×	×	—	—
Dezincification of brass	×	—	×	×	—	—	×	×
Ammonia attack of brass	—	—	×	×	—	—	—	—
Dissimilar metal corrosion	×	×	×	×	—	—	×	×
Chloride attack of stainless steel	×	—	×	—	—	—	×	—
Caustic attack	—	—	—	—	—	×	—	—
Hydrogen damage	—	—	—	—	—	×	—	—
Formation of inorganic deposits	×	×	×	×	×	×	×	×
Microbiological fouling	×	—	×	×	—	—	×	—
Deterioration of cooling tower lumber	—	—	×	—	—	—	—	—
Carryover	—	—	—	—	×	×	—	—

Key:

A. Once through cooling water systems
B. Closed circulating cooling water systems
C. Open circulating cooling water systems
D. Air conditioning systems
E. Low pressure steam generating systems
F. High pressure steam generating systems
G. Process water
H. Potable water

Courtesy of Drew Industrial Division, Ashland Chemical, Inc. Subsidiary of Ashland Oil, Inc.

Table 1-3. Treatment of Water from Rivers, Ponds and Wells

Raw Water Treatment may have to include two or more of the following steps:

1.	Aeration:	Water falling through air (spray, cascade, forced draft air), air bubbling through water (air diffusion method).
2.	Subsidence:	Clarification by settling methods; removal of bacteria.
3.	Coagulation:	Agglomerate suspended particles to make water readily filtered and to settle out coagulated matter prior to filtration.
4.	Sand Filtration:	To remove suspended solids and turbidity.
5.	Precoat Filtration:	Diatomaceous earth filtration—oil removal.
6.	Chlorination:	Hypo and dechlorination (when sewage waters have to be used).
7.	Hydrogen Sulfide:	Removal to reduce corrosive character of certain waters.
8.	Iron Removal:	By cation exchange, coagulation, sedimentation, oxidation and filtration as water and plant conditions dictate.
9.	Lime-Soda Softening (Cold-Process):	Calcium, magnesium salts, turbidity, iron and free CO_2 removal.
10.	Phosphate Softening (Hot Process):	Calcium, magnesium, silica and turbidity removal.
11.	Silica Removal (Hot Process):	Removal of soluble silica by magnesium compounds from water to prevent formations of calcium, magnesium and sodium-alumino silicate scales.
12.	Sodium Zeolite Softening (Cation Exchange) Cold (50°F) to Hot (180°F) Process:	Hardness salts of calcium and magnesium removal. Non-turbid water removal of corrosive gases by deaeration or using corrosion inhibitors.
13.	Hydrogen Zeolite Softening (Cation Exchange):	Removal of calcium magnesium bicarbonate; not very effective for water of high hardness cation exchange and low alkalinity or high in silica.
14.	Demineralization (Deionization):	Removal of mineral salts by ion exchange and removal of silica with strongly basic anion exchanger.
15.	Dealkalization by Chloride-Anion Exchange:	To prevent high boiler water alkalinity leading to foaming and solids carryover with the steam and to reduce alkalinity (sodium zeolite softener plus chloride anion exchanger).
16.	Hot Lime—Hot Ion Exchange Softening (Hot Lime-Sode-Hot Phosphate):	To reduce hardness, alkalinity, total solids and silica. (Generally requires sodium zeolite afterwards).

| | PURITY LEVEL REQUIREMENTS | | |
	Low (10,000-1M)	High (1M-9M)	Ultra High (10M-18M)
Semiconductor			
semiconductor manufacturing			●
Pharmaceutical & Cosmetic			
pretreatment of stills		●	
laboratory water		●	
animal cage washing	●		
Institutional			
quantitative analysis			●
glassware washing	●		
general laboratory	●	●	●
Chemical Process			
product solvents & solutions		●	
hydrolysis	●		
cooling water	●		
Plating/Metal Finishing			
jewelry, dishware	●		
final rinse water	●		
printed circuit mfg.		●	
electrical parts		●	
Boiler Feed Water	●		

Figure 1-6. Purity Level Required for Various Processes

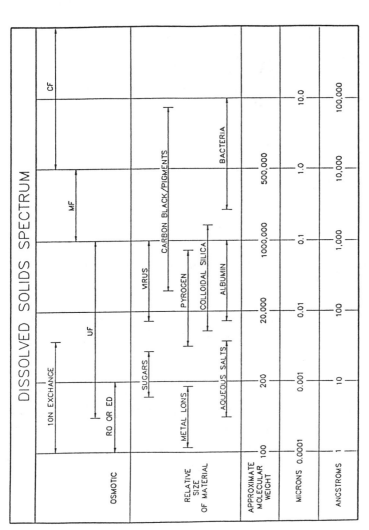

Figure 1-7. Molecular/Atomic Spectrum of Water Impurities

Note the type of equipment used for removal of dissolved and particulate solids. The chart does not show chemical means of removal.

2

Water Treatment
Program Management

Because of variations in the quality of source water and differing requirements for water quality, no single manual can address every eventuality. This chapter begins with a brief presentation of information on the importance of water treatment in boiler and cooling water systems. The balance of the chapter deals with how to manage (plan, organize, execute, and control) a high-quality water treatment program. Guidelines presented for managing a water treatment program are applicable to most industrial activities. Modifications may be necessary to adapt the program to the needs of a particular installation.

Water is often called "the universal solvent" because of its ability to dissolve, to some extent, practically everything. This characteristic is especially important to process water users and plant operators. While additives, either dissolved or suspended, can improve water quality for some uses, in most instances impurities cause corrosion, fouling, inefficient operation, and eventual failure of plant equipment. With increased boiler pressure, higher quality feedwater is required. Timely detection and correction of parameter abnormalities or treatment deficiencies are essential to managing and maintaining an effective treatment program.

BOILER SYSTEMS

The objective of boiler feedwater treatment is control of boiler water chemistry. With this control, scaling, corrosion, carryover, and embrittlement can be either reduced or eliminated. Feedwater treatment consists of both limiting the introduction of contaminants into the boiler and introducing internal treatment chemicals to maintain water chemistry at appropriate levels. Boiler feedwater may range from all makeup to all returned condensate, with treatment requirements varying between these extremes. Regardless of the source or quality of the boiler feedwater, some solids will pass into the boiler. Therefore, the levels of treatment chemical residuals are important control parameters that must be monitored.

Table 2-1 displays problems caused by specific impurities and suggested means of treatment where water is to be used in boilers. Many of these problems also occur when these contaminants are present in water used for other purposes.

The effects of inadequate or improper water conditioning in boiler apparatus are shown by Table 2-2.

The accumulation of solids in a boiler is controlled by blowdown. Blowdown consists of replacement of drained boiler water, with its accumulated solids, and replacing it with treated feedwater. Dissolved solids tend to concentrate near the water surface in the steam drum. Surface blowdown is therefore most effective in reducing the concentration of dissolved solids. Bottom blowdown is employed to remove precipitated sludge from the boiler "mud drum." Blowdown, however, results in the loss of heated water and treatment chemicals.

Economical operation requires careful control of blowdown to maintain safe solids levels, while minimizing both heat and chemical additive losses.

The condensate return system is a revealing point for monitoring total system performance. The amount and nature of condensate contaminants often will identify problems of both carryover and corrosion and will suggest corrective action. Identification of

Table 2-1. Common Impurities in Water and Possible Effect When Used for Boiler-Feed Purposes

Constituent	Chemical formula	Principal source of contaminating substance	Possible effect when present in boiler waters	Treatment for removal from water
Suspended solids	. . . .	Surface drainage, industrial wastes	Priming, foaming, sludge, or scale	Plain subsidence, coagulation, filtration, evaporation
Silica	SiO_2	Mineral deposits	Scale	Plain subsidence, coagulation, filtration, evaporation, ion exchange
Calcium carbonate	$CaCO_3$	Mineral deposits	Scale	Softening by chemicals, ion-exchange materials, evaporators
Calcium bicarbonate . . .	$Ca(HCO_3)_2$	Mineral deposits	Scale	Softening by heaters, chemicals, ion-exchange materials, evaporators
Calcium sulfate	$CaSO_4$	Mineral deposits	Scale, corrosion	Softening by chemicals, ion-exchange materials, evaporators
Calcium chloride	$CaCl_2$	Mineral deposits	Scale	Softening by chemicals, ion-exchange materials, evaporators
Magnesium carbonate . .	$MgCO_3$	Mineral deposits	Scale	Softening by chemicals, ion-exchange materials, evaporators
Magnesium bicarbonate .	$Mg(HCO_3)_2$	Mineral deposits	Scale	Softening by chemicals, ion-exchange materials, evaporators

Magnesium chloride ...	$MgCl_2$	Mineral deposits	Scale, corrosion	Softening by chemicals, ion-exchange materials, evaporators
Free acids	HCl, H_2SO_4	Mine drainage, industrial wastes	Corrosion	Neutralizing, followed at times by softening or evaporation
Sodium chloride	$NaCl$	Sewage, industrial wastes, mineral deposits	Inert, but may be corrosive under some conditions	Evaporation and demineralization by ion-exchange materials
Sodium carbonate	Na_2CO_3	Mineral deposits	Priming, foaming, embrittlement	Evaporation and demineralization by ion-exchange materials
Sodium bicarbonate ...	$NaHCO_2$	Mineral deposits	Priming, foaming, embrittlement	Evaporation and demineralization by ion-exchange materials
Carbon acid	H_2CO_3	Absorption from the atmosphere, mineral deposits, decomposition of organic matter	Corrosion	De-aeration
Oxygen	O_2	Absorption from the atmosphere	Corrosion	De-aeration
Grease and oil	....	Industrial wastes	Corrosion, deposits, priming, foaming	Coagulation, filtration, evaporation
Organic matter and sewage	....	Domestic and industrial wastes	Corrosion, deposits, priming, foaming	Coagulation, filtration, evaporation

Source: Water Conditioning for Industry, Shepard Powell

Table 2-2. Effects of Inadequate or Improper Water Conditioning

Effect	Constituent or Result	Remarks
Scale.	Silica.	Forms a hard, glassy coating on internal surfaces of boiler. Vaporizes in high pressure boilers and deposits on turbine blades.
	Hardness.	$CaSO_4$, $MgSO_3$, $CaCO_3$, and $MgCO_3$ form scale on boiler tubes.
Corrosion.	Oxygen.	Causes pitting of metal in boilers, and steam and condensate piping.
	Carbon dioxide.	Major causes of deterioration of condensate return lines.
	$O_2 + CO_2$.	Combination is more corrosive than either by itself.
Carryover.	High boiler water concentrations.	Causes foaming and priming of boiler and carryover in steam, resulting in deposits on turbine blades and valve seats.
Caustic embrittlement.	High caustic concentration.	Causes intercrystalline cracking of boiler metal.
Economic losses.	Repair of boilers.	Repair pitted boilers and clean heavily scaled boilers.
	Outages.	Reduce efficiency and capacity of plant.
	Reduced heat transfer.	High fuel bills.

contaminant sources and control of treatment systems depend on careful monitoring of key parameters. For example, excessive hardness, conductivity, and turbidity can indicate in-leakage through heat exchangers and condensers; iron, copper, and dissolved oxygen indicate corrosion; and either sodium or silica in condensate can indicate carryover.

COOLING WATER SYSTEMS

Cooling water systems are used to remove heat from engines, air compressors, refrigeration condensers, steam power systems, and industrial processes. These cooling systems are classified as either once-through or recirculating. Once-through systems often require nothing more than chlorination to prevent biological fouling of heat exchangers. Treatment is more critical in open recirculating systems because of solids build-up due to evaporation. As hardness and other solids increase, probability of mineral scale formation in heat exchangers increases. To combat scale damage, chemical additives are used to keep scale-forming salts in solution.

Corrosion in cooling tower systems can be general or localized. Dissolved oxygen and carbon dioxide are prime corrosion developers. Corrosion control is provided by addition of inhibitors. In general, chemical attack can be controlled by maintaining constant pH and appropriate levels of alkalinity and inhibitor residual. Slime accumulation and fouling may be prevented by the addition of chlorine or other biocides. Excess chlorine concentration in conjunction with alkalinity in water, however, can be destructive to wood in the system. Solids concentration is controlled by blowdown. In all instances, monitoring for hardness, pH, inhibitors, chlorine, other chemical additives, and solids is essential for cost-effective treatment. Disposal of blowdown may constitute a significant ecological or legal problem. This may radically affect the form of treatment to be selected.

ULTRA PURE WATER SYSTEMS

The need for water that approaches the theoretical maximum level of purity, with an electrical resistance of about 28 million ohms per centimeter, is well established. Nuclear, medical, and electronic processes need such purity for successful operation. The techniques required to produce or purify water from miscellaneous sources to this level of purity will be described herein. Starting with simple sand filtration and evolving to more complex methods, water from any source can be brought into compliance with the most stringent requirements.

PLANNING THE PROGRAM

In planning a water treatment program, it is first necessary to establish facility water quantity and quality requirements. This assessment should include an inventory of the number and types of systems that require high-quality water. Also required is a determination of purity specifications, existing water quality, and equipment condition for each application. Applicable standards should be reviewed. Typical standards, for example, are included in publications by the American Boiler Manufacturer's Association, the American Society of Mechanical Engineers, and by equipment manufacturers in their suggested operating procedures.

Systems to be surveyed include large central steam and power plants, smaller heating boilers, cooling tower systems, engines, and air compressors.

For each system or system element, the following information should be gathered and evaluated in the planning phase:

- Quantity of water required: volume and rate.
- Quality of water required: pH, conductivity or TDS, and limits on hardness, silica, dissolved oxygen, etc.
- Instrumentation required.
- Type and condition of existing instrumentation and evidence of scale, corrosion, and deposits in system components.

Water quality requirements for specific applications are discussed in Chapter 5. Based on the information gathered, requirements in terms of staff, equipment, and procedures can be developed.

ORGANIZING THE PROGRAM

With water treatment requirements defined for an installation, personnel must be organized to implement the program. The organization process consists of identifying an appropriate staff, training the staff to execute the program, and integrating the program into overall installation management.

TRAINING OPERATORS

Adequate operator training is essential to achieve an effective water treatment program. Operators must be qualified to understand not only the operation of their specific plan, but the reasons for specified procedures. An understanding of the reasons for and the results of chemical treatment is essential. Knowledge of thermodynamic and mechanical interactions of plant equipment is important. The operator must be particularly aware of safe operating procedures.

Since internal chemical treatment of system water is a significant factor in achieving reliable and efficient plant operation, operators should have an understanding of the subject. Specifics of water treatment where training should be provided are:

- the necessity for pretreatment: water quality and associated problems
- methods of pretreatment: chemical addition, ion exchange, reverse osmosis, distillation, electrodialysis
- deaeration and oxygen scavenging
- the necessity for internal boiler water treatment and maintenance of concentration residuals
- causes and effects of deposits and scale
- causes and effects of corrosion
- types and costs of chemicals available for internal treatment of boiler water, such as phosphate polymers, sulfite, and hydrazine
- application of treatment to boiler water: where and how to feed

- blowdown: surface vs. bottom, internal provisions for blowdown, estimating blowdown, continuous vs. intermittent, blowdown controllers, cycles of concentration
- blowdown heat recovery: flash tanks, heat exchangers, system arrangement
- condensate line corrosion: iron and copper oxide in returns, filming and neutralizing amines, effect of hydrazine
- steam quality: carryover, ABMA and ASME standards, effects of carryover and silica, steam purity
- boiler lay-up procedures
- once-through cooling water systems: need for and means of maintaining clean condensers
- recirculating cooling water systems: water treatment methods.

Training should be provided in steps that are geared to the capabilities of the operators. Weekly sessions, with handouts, periodic reviews, and testing, should be planned to assess both trainees' retention and understanding of material and effectiveness of teaching methods.

EXECUTION

An effective water treatment program requires establishment of an inspection and monitoring program, and development of a reporting system to keep the manager informed on the status of both water quality and water treatment programs.

Detection and correction of either abnormalities or inadequate treatment are essential to managing and maintaining an effective program. This is accomplished by both routine inspection and frequent, consistent monitoring. Instrumentation and laboratory analysis equipment should be checked for specified operation on a regular basis.

STEAM PLANT MONITORING

In continuously manned plants, operators and laboratory personnel should monitor and log:

- *Effluent from the external treatment system.* Every four hours hardness, conductivity, and silica should be monitored to ensure operation within specification limits. Excursions from acceptable limits indicate either improper operation or equipment failure.

- *Deaerator effluent.* Dissolved oxygen and water temperature should be checked daily to ensure that residual oxygen is less than the established limit (usually 0.005 mg/1) and that water temperature is close to steam temperature in the deaerator. If parameters are not as required, check deaerator venting, steam supply, and internals for either plugged spray nozzles or corroded trays.

- *Boiler water.*

  Conductivity: Check boiler water every four hours to determine if conductivity (TDS) is within specified operating range. Increase or decrease continuous blowdown rate to achieve the specified reading.

  pH: Check every four hours. If not within specified range, adjust either chemical feed or other controlling factors.

  Oxygen scavenger: Check by chemical test once each watch. The residual level should conform with recommendations of the scavenger supplier. Adjust for this residual, when required, by modifying the stroke of the feeder pump, or by altering the concentration in the feed tank. Never modify both at the same time.

  Total alkalinity/causticity: Check by chemical test once each watch. Caustic feed should be adjusted, as required, to maintain the recommended residual level.

Phosphate or polymer: Check by chemical test once each watch. Adjust chemical feeders, as required, to maintain the recommended residual level.

Silica: Check daily for the level of silica in boiler water. If over the specified limit, blowdown should be increased. A check should be made for the source of silica intrusion.

- *Condensate.* Returned condensate should be checked at least daily for conductivity, metallic contamination, and hardness. If an increase over prescribed limits is observed, a complete laboratory analysis should be performed to determine the source of contamination. Analysis should include iron and copper determinations. When significant impurities are detected, condensate should either be diverted to another use or discarded until the source of contamination has been located and eliminated.

RECIRCULATING COOLING WATER MONITORING

Cooling tower basin water should be controlled to maintain water quality at optimum conditions with minimum fouling, scaling, and deterioration by monitoring the following, by plant operators or roving patrols:

- *Conductivity (total solids).* Depending on makeup supply water and system operating parameters, basin water solids should be controlled to a range of one to four cycles of concentration. That is, solids concentration in the basin should be about five times that of the makeup supply water. This parameter varies with makeup water quality. Basin water solids (conductivity) should be checked weekly and continuous blowdown adjusted to maintain a desired level of solids concentration.

- *pH control.* Control of basin water pH is essential to minimize deposition of scale and to reduce corrosion. pH should be checked weekly and acid feed adjusted accordingly to hold alkalinity within acceptable limits.

- *Corrosion inhibitor.* Since corrosion inhibitor is lost in blow-down, the feed rate of inhibitor should be checked and adjusted whenever the basin blowdown rate is adjusted. Chemical dispersant should be added with inhibitor to precipitate and fluidize some solid constituents in the basin water.

- *Biocide.* Biological growth occurs in many cooling towers and basins. Concentration of basin water biocides that inhibit biological growths, should be checked weekly. Biocide feed rate should be adjusted to pace makeup supply water rate. If ozone is used, the residual in the basin should be checked daily, and feed adjusted accordingly.

- *Maintenance.* Personnel should inspect daily for operation of automatic chemical feed equipment; renew chemicals; and send water samples, weekly, to the laboratory for analysis. Personnel should make weekly adjustments, based on laboratory analyses. Prompt repair and maintenance should be provided for automatic feed and control equipment and a complete record should be maintained for each tower.

INSPECTIONS

Routine periodic inspections should be conducted, in addition to daily monitoring. For cooling towers, quarterly inspections should be performed, and can be made when the tower is operating. Scaling, slime, algae, and wood fill deterioration are readily visible conditions. For boilers, inspection is less frequent, since a boiler must be shut down to perform the inspection. Inspections are normally annual, but every major shutdown should be regarded as an inspection opportunity. Scale samples should be taken from the sludge in the boiler drum and scraped from the tubes for analysis. The thickness of scale on or in the tubes should be estimated and reported.

Water treatment affects corrosion and scaling on the water side of heat exchangers, coolers, and refrigeration condensers. Equipment should be inspected and evaluated in the same manner as for boilers.

REPORTS

Managers should receive a monthly summary of inspection and monitoring program data. A sample monthly report is provided in the Appendix. For a large continuously manned central boiler plant, the report should contain results of daily tests, in graph form with the following data:

- Conductivity (TDS)
- pH
- Hardness
- Dissolved oxygen
- Alkalinity
- Silica (for high pressure systems)

The report should also include the following:

- Fuel consumed per pound of steam
- Pounds of steam produced
- Make-up water used
- Condensate returned
- Hours of operation
- Condensate quality
- Blowdown rate
- Chemical consumption
- Softener or demineralizer usage

The summary report should include on each graph the allowable limits of the respective parameters, facilitating quick assessment of the acceptability and trends of reported results.

For smaller plants and low pressure boilers, the monthly report should include:

- Hardness

- Dissolved oxygen
- TDS (conductivity)
- pH
- Chemical residuals
- Alkalinity
- Steam produced, makeup water used, and condensate returned
- Chemicals consumed

Similarly, cooling tower reports should provide the following data:

- Conductivity
- pH
- Total dissolved solids
- Corrosion inhibitor residual level
- Biocide residual content
- Blowdown rate
- Chemical residuals

Managers should receive all routing inspection reports. Equipment condition reports indicate the adequacy of a water treatment program. Inspections should be conducted on the following schedule.

System	Minimum Frequency
Boilers	Annually or on every major shutdown
Cooling Towers	Quarterly
Heat Exchangers	Same as boilers
Condensers	Same as boilers

CONTROLLING THE PROGRAM

Once a water treatment program has been implemented, it is necessary to continuously review results to determine if it is meeting requirements. Both requirements and standards were developed in the planning phase. Examples of standards for boiler plants are provided in Tables 5-1 and 5-2 in Chapter 5. The manager must be sensitive to any change in requirements that necessitates modification to the program.

CHECKING AND APPRAISING THE PROGRAM

Following are parameters to be checked in evaluating the adequacy of a water treatment program and some hints as to the source of possible problems:

Makeup Water

- Adequacy of supply
- Analysis of source water—comparison with quality required
- Seasonal variations in Source Water quality
- Adequacy of treatment system capacity
- Analysis of treatment systems effluent—comparison with process requirements; seasonal variation
- Treatment system operation—labor required; adequacy of instrumentation and controls; visible evidence of component corrosion, deterioration, or malfunction

Boiler Systems

- Evidence of deposits, scaling, or corrosion in the system—analysis of deposits
- Adequacy of deaerator operation—suitable vent plume, water temperature, residual oxygen content, residual oxygen scavenger
- Total dissolved solids (conductivity) within prescribed limits; too low a reading indicative of excessive blowdown; wasteful of heat, water, and chemicals (refer to Tables 5-1 and 5-2 for limits)
- Chemical residuals—phosphate, oxygen scavenger, and other chemicals within limits recommended by supplier; too high residual content indicative of faulty, wasteful feed rate; too low, inadequate protection
- pH—between 9 and 11 (corrosion rate should be considered)

Steam Condensate Systems

- Evidence of corrosion, particularly where condensate forms or flows—indicative of carbon dioxide and/or oxygen in system, the result of faulty water treatment and/or infiltration into condensate returns
- Excessive carryover—check for deposits in saturated steam header; check steam quality by calorimeter or ion detector; usually results from excessive solids in boiler water or generally high boiler water level
- Iron, copper, or hardness in condensate—indicative of corrosion in system or infiltration from external sources

Cooling Water Systems

- Visible condition of cooling tower and basin: algae, slime, or mold—indicative of insufficient biocide; metal corrosion—insufficient inhibitor; deterioration of wood members—excessive chlorine or alkalinity; deposits or scale—excessive solids (insufficient blowdown) or inadequate chemical treatment
- Waterside of heat exchangers: scale, deposits, or corrosion—evidence of insufficient chemical treatment; fouling—insufficient biocide
- Once-through systems—check for visible indications of fouling, plugging, and organisms

Treatment Costs

- Cost per unit for treated makeup water as compared to other similar treatment systems
- Costs for internal treatment chemicals per unit of production, compared with costs for similar systems

Operating Records and Reports

- Maintenance of complete and accurate operating logs of water treatment systems
- Maintenance of accurate records of chemicals consumption, plant loads, chemical costs
- Reliable recording of maintenance operations and costs

CORRECTIVE ACTION

The preceding checklists will aid in assessing the adequacy of treatment practices, in identifying problems, and in developing an action plan. More detailed information to assist in resolving problems is provided in the following chapters.

Assuming that the treatment program is followed, and that monitoring and inspection reports indicate that treatment is ineffective, the following steps should be taken:

- Review supply water analyses. Determine if significant changes have occurred since initiation of the treatment program.

- Review the adequacy of performance parameters for the treatment system, such as capacity and quality of source water and of effluent.

- Determine if significant changes have occurred in plant operating conditions.

- Review performance of personnel who make tests and adjustments.

- Consult with the on-site water treatment specialist regarding additions to or changes in treatment practices.

- Try to determine why treatment is ineffective: competence of personnel, chemicals, dosages, changed conditions.

- Adjust program as necessary, but change only one parameter at a time. Results should be monitored closely to determine the changes that have the most beneficial effect.

- In the case of gross problems, an entire treatment program may need modification. If this is planned, all equipment, boilers, heat exchangers, and cooling towers should be descaled, cleaned and repaired. This allows evaluation of a modified treatment program.

Alternatives available are to hire a consultant, or a chemical supply company specializing in water treatment, or to contract for water treatment services. The selection of one or more of these courses of action depends on available financial resources and the magnitude of the problem.

Water treatment chemical supply companies are situated in the metropolitan areas of all large cities in the U.S. In addition, there are nationwide organizations with branch offices and distribution facilities in all major metropolitan areas. These are the companies who have developed the art of internal chemical water treatment.

Most companies issue bulletins describing water treatment products and services offered. The product listings are especially informative.

When dealing with a specialist company, one should be aware that each firm promotes only its own products. One should not hesitate, however, to ask questions regarding competitive products. Some products and formulations are uniquely suited to specific problems. Furthermore, water treatment services should be reviewed with respect to what competing companies may offer. Above all else, services of the company engaged should be judged on the basis of results achieved; adequacy should be assessed, including the cost of the program.

One may choose to contract for all or a portion of the services required. The contracting decision will be affected by factors such as the size and complexity of the plant or plants to be operated and maintained, the available pool of labor, and the skills and level of training and education of personnel.

At the very least, procurement of internal treatment chemicals should include arrangements for periodic visits and tests by the supplier's specialists to determine and recommend dosage rates and residual concentrations to be maintained.

A further step is to contract for services of a specialist consultant to provide management oversight of a water treatment program. Specifications for contracting for this type of service could stipulate the following:

A thorough check of plant operating conditions shall be made each month. Tests of untreated supply water and cooling towers shall be made, as required, during regular monthly visits.

(a) At the time of each monthly service visit, the service engineer shall review the records of control tests of the steam generating plants, discuss these with plant personnel, and offer comments or constructive criticism. The service engineer shall conduct tests and inspections needed to check the accuracy of data recorded. After each service visit, and before the engineer leaves the station, recommendations shall be submitted, in writing.

(b) Report forms shall be furnished by the service contractor for recording, in duplicate, results of daily tests performed by plant personnel. Constructive criticism of these reports shall be submitted, in writing, monthly. Daily reports shall be reviewed promptly and if the results indicate the need for revisions in control procedures, recommendations and/or instructions shall be issued immediately, in writing.

(c) During the period of each monthly service, samples of water shall be taken from specified points in the plant systems for analysis by the service contractor. Analysis of samples constitute an additional check on the accuracy of plant tests. Analyses, together with the plant control tests and the reports of the service engineer, serve as a basis for a complete review of plant conditions by the service contractor.

(d) Tests of untreated supply water that constitutes feedstock for boiler feedwater and control tests on cooling towers shall be performed as required.

(e) In addition, the service engineer shall be responsible for instructing plant operating personnel in test methods. The service engineer shall assist in inspection of internal boiler surfaces and feedwater equipment; such inspections shall normally be conducted once during the contract period, unless an unusual condition should develop that requires more frequent inspection. Inspection is for the purpose of providing consulting advice regarding conditions found by the service contractor.

Finally, the ultimate step is to contract for complete treated water supply and water treatment management. This could include installation and operation of a water treatment facility, either fixed or trailer mounted, or the supply of treated water by tank transport for lesser makeup water requirements. In these cases the contractor could also provide all services required for testing and maintaining the internal chemical treatment systems.

WATER TREATMENT DEVICES

Water treatment devices or gadgets purport to end all water problems, and to replace all chemical treatment. Promoters claim that these devices use magnetics, ultrasonics, or electrostatics, to control scale, corrosion, and algae in boiler and cooling tower water systems without the use of chemicals and with little or no expenditure of energy. These devices are discussed in detail in a subsequent chapter.

BENEFITS OF EFFECTIVE TREATMENT

A well-managed water treatment program will minimize scale and fouling buildup from organic and inorganic sources and significantly reduce corrosion potential. Maintenance of good quality water will result in the reduction of:

- equipment downtime and operating and repair costs,
- energy consumption through higher heat transfer efficiency and less energy loss in boiler blowdown, and
- environmental pollution by decreasing harmful effluent.

As an example of one benefit, the following information has been extracted from a recent energy conservation study that shows operational improvements available through improved water treatment.

Application	Estimated Annual Energy Savings	Cost to Implement
Central Heating and Power Plant Boilers	$2,560,000	$3,000,000
Central Plant Condensers	$ 916,000	$1,850,000
2 Low Pressure Heating Hotel Steam Boilers	$2,170,000	$7,595,000
17 Air-Conditioning and Refrigeration Cooling Towers	$ 723,000	$ 958,000

Additional savings are achieved through reduction of boiler shutdown and cleanup requirements made possible by an improved water treatment program.

3

Water Quality

Few engineering or industrial operations can be performed without a conditioned water supply.

Pure water is tasteless, colorless, and odorless. Water molecules are composed of two atoms of hydrogen and one atom of oxygen. Water possesses several unique properties, among which are a tremendous capacity to absorb and store heat and the ability to dissolve everything, to some degree. Because of this solvent property, water contains impurities. These impurities are the source of scale that can deposit in pipe, boiler tubes, and other surfaces that come in contact with water. Dissolved oxygen, the principal gas present in water, is responsible for corrosion of metals. Figures 3-1 and 3-2 illustrate the effects of oxygen corrosion and scale buildup on water piping.

SOURCES OF WATER

The hydrologic cycle begins with water evaporating from land mass areas and the oceans. This moisture precipitates in the form of rain, snow, sleet, or hail. As it falls, it comes in contact with gases in the atmosphere (naturally occurring as well as pollutants) and suspended particles such as dust and smoke. By the time the precipitation reaches the Earth's surface, it contains dissolved gases and minerals collected from the atmosphere. Studies show that these impurities can result in precipitation with a pH range of 3.5 (highly acidic) to 7.1 (slightly alkaline). Rain water is not pure.

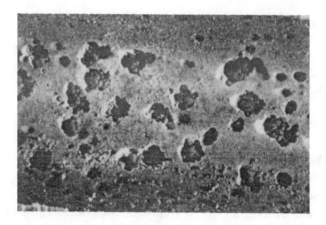

Dissolved oxygen causes attack in the form of broad pits. Active pits contain reduced black oxide along the concave surface of the pit with the surrounding area covered with red ferric oxide.

Figure 3-1. Corrosive Attack Caused by Dissolved Oxygen

Figure 3-2. Scale Deposits in Water Line

Most often, fresh water is used for industrial applications. The two most important sources are surface water and ground water. A portion of the runoff from rain or melting snow soaks into the ground. Part of it collects in ponds and lakes or runs off into creeks and rivers. This portion is termed surface water. As water flows across the land surface, minerals are dissolved and the force of the flowing water carries finely divided particles and organic matter in suspension. Both the character of the terrain and the geology of an area influence the quantity of impurities found in surface water.

The portion of water that seeps into the Earth's crust and in natural reservoirs called "aquifer" is called "ground water." Underground supplies of fresh water differ from surface supplies in two important aspects, which make ground water superior as an industrial water source: relatively constant temperature and absence of suspended matter. Ground water, however, tends to be higher in dissolved mineral content than surface supplies.

Potable utility provided water also is used as a fresh water source. This water, typically obtained from a municipal facility, is clarified, pH adjusted, and treated to assure biological purity. Note that although this water may be pure enough to drink, it usually requires treatment before use in industrial processes. Table 3-1 shows various categories of potable water and the chemical effects when in contact with piping.

In many areas, water supplies are limited. Facilities requiring large amounts of process and cooling water find it necessary to conserve available supplies by recycling. Use of effluent streams from sewage treatment plants is a common practice.

The use of sea water is limited chiefly to cooling applications. Facilities located in coastal zones that have large cooling water requirements frequently use sea water. Users should be aware of restrictions on thermal pollution.

Table 3-1. Characteristics of Water in Closed Systems

Potable water categorization is applicable to all metals for scaling but only to iron, steel (not stainless) and galvanized steel for corrosiveness.

Category	Calcium (Ca), ppm	Sulfate (SO₄), ppm	Silica (SiO₂), ppm	Dissolved oxygen, ppm	Character
1A	—	As found	0-15	1-10	Extreme corrosion hot and cold.
1B	0-18	0-25	0-15	0-1	Moderate corrosion hot and cold. Extreme corrosion with CO_2 > 8 ppm.
1C	—	0-60	>15	1-5	Slight corrosion cold, considerable hot. Aggressiveness reduced and perhaps not troublesome due to high natural SiO_2.
2A	—	>Ca but not <25	0-15*	1-10	Considerable corrosion hot, moderate cold. May be slightly scale forming very hot.
2B	18-35	0-25	0-15	1-10	Moderate to slight corrosion hot and cold. May be scale forming hot.
2C	18-35	<Ca	>15	1-8	Corrosion unlikely. May be scale forming hot.
2D	18-35	<Ca	As found	0-1	Corrosion unlikely. May be scale forming hot.
3A	35-75	<1½Ca	0-15	1-10	Moderate corrosion hot, slight cold. Considerable scale formation hot.
3B	35-75	>1½Ca	0-15	1-10	Considerable corrosion hot, slight cold. Considerable scale formation hot.
3C	35-75	<1½Ca	>15	1-10	Considerable scale formation. Slight corrosion hot.
3D	35-75	As found	As found	0-1	Considerable scale formation. Corrosion unlikely.
3E	35-75	>1½Ca but <3Ca	>30	1-10	Corrosion unlikely hot and cold. Excessive scale formation.
4A	>75	<2Ca	0-30	1-10	Excessive scale formation. Corrosion unlikely to slight cold, slight to moderate hot.
4B	>75	>2Ca	0-30	1-10	Excessive scale formation. Galvanic corrosion considerable hot and cold.
4C	>75	<3Ca	>30	1-10	Excessive scale formation. Corrosion unlikely.

*With SiO_2 over 15 ppm, corrosion may be reduced in proportion to SiO_2 content.

Notes: Presence of chlorides in concentrations greater than 100 ppm, with high sulfates, renders a water more corrosive than indicated by category above. Presence of carbon dioxide in concentrations exceeding 5 ppm accelerates corrosion processes where category groups indicate corrosion. In concentrations exceeding 20 ppm, it may cause an indicated noncorrosive water to be corrosive.

Terms: Extreme or excessive — where effects necessitate immediate corrective action.
Considerable — where corrective action is desirable.
Moderate — where corrective action is questionable and depends on economy effected.
Slight — where effect is too slight to warrant corrective action.
Unlikely — where effects are possible but not probable.

Source: Myers and Obrecht, "Potable Water Systems in Buildings: Deposit and Corrosion Problems." *Heating/Piping/Air Conditioning,* May 1973, pp. 77-83.

IMPURITIES IN WATER

Natural water, always more than a simple combination of hydrogen and oxygen, contains impurities. The Mississippi River, for example, carries about 400 million tons of suspended material per year to the Gulf of Mexico and approximately 0.025 percent dissolved material. The water in Great Salt Lake contains about 23 percent dissolved material, compared to 3.5 percent sea water.

Solid Impurities

An analytical examination of natural water will show that it contains many materials. Solids in water are either suspended or dissolved.

Suspended solids such as sand and silt, known as turbidity, can be removed from water by settling, when the water can be allowed to stand quietly for a period of time. Since settling is seldom practical, most suspended solids are removed by filtration.

Dissolved solids cannot be removed by simple filtration. They can, however, be removed by various types of membranous molecular sieve filtration processes, or by ionic processes. The chart in Figure 1-7 shows the range of solids contained in water, relative particle sizes, and methods of removal by membranes or conventional filtration. These materials are mostly calcium, magnesium, and sodium cations, and chloride, bicarbonate, and sulfate anions. Other inorganic soluble solids such as silicates, nitrates, and nitrites are found in lesser amounts. Organic substances may also be dissolved in water. These may be substances leached from natural debris, or introduced through industrial waste and sewage.

A process called ionization, an important feature of the solution process, should be understood in order to grasp some of the techniques of water treatment and the chemistry of solutions.

Every chemical compound—and chemical compounds make up the bulk of the world as we know it—is composed of molecules, the smallest particles of matter that retain the properties of compounds.

The molecules of a large number of these compounds, in turn, are composed of still smaller particles, known as ions, actually atoms bound together by electrical charges. When such compounds are dissolved in water or certain other solvents, they become "ionized," that is, their ions are free to move about and take part in chemical reactions. Those ions that bear a positive charge are called cations. Ions that bear a negative charge are called anions. The sodium ion of sodium chloride, for instance, is a cation. The chloride portion of the same compound is an anion. Of course, not all compounds dissolved in water are ionized in the same fashion. Some, like sugar or alcohol, are negligibly ionic. Such compounds are not capable of ion exchange Others, like water itself, ionize very slightly or insignificantly. Silica, in solution as SiO_2, is one such compound. It is important to know this, because it is a major problem in many water uses. Only those materials that yield ions in solution are subject to ion exchange processes.

Calcium and magnesium salts contribute to water hardness. Calcium salts, usually present as bicarbonate, change to calcium carbonate when water is heated. This forms scale on heat transfer surfaces and reduces efficiency. Calcium sulfate is also present in most natural waters and is another source of scale. Magnesium salts are less important scale formers.

Nature's primary carbonate mineral, calcium carbonate, is only slightly soluble at 13 mg/liter but in water saturated with carbon dioxide the solubility increases about thirtyfold. This unusual change is due to the formation of the bicarbonates or acid carbonate ion. The development of soluble calcium, magnesium, and barium bicarbonate by carbon dioxide charge waters is responsible for the many types of cave minerals, building stones such as marble or travertine, and the hardness of raw water supplies.

Alkalinity, the acid neutralizing power of water, is associated with basic cations. These include sodium, calcium, and magnesium in combination with bicarbonate, carbonate, and hydroxyl anions. While all three of these anions can theoretically exist simultaneously, only bicarbonate is present in natural water and open recirculating systems where the pH is below 8.3.

The chloride ion is one of several anions found in water and is frequently used to determine the concentration cycles of cooling water. The ratio of chlorides in tower water to the chlorides in make-up water is equal to the system concentration cycle, unless heavy chlorination is practiced.

Sulfate ions are common constituents of natural water because of the high incidence of combined sulfur in rock. Runoff water from mines and industry increases sulfate concentration in surface waters. In industrial cooling water, calcium and sulfate ions are generally maintained at levels below the point of calcium sulfate (gypsum) scale formation. Waters drawn from limestone formations are high in calcium carbonate hardness and can cause calcium sulfate scale if the water is treated with sulfuric acid for pH adjustment. Close control may be required when treating with sulfuric acid to prevent sulfate scale formation.

When iron is present in cooling water, it may be a natural constituent, a corrosion product, or a combination of both. Iron occurs as ferrous bicarbonate in well water and precipitates in the insoluble ferric form when aerated. Deposits resulting from this precipitation are recognized by a characteristic red color. Certain bacteria use iron in metabolic processes and cause trouble by producing iron deposits.

Manganese does not usually cause problems, though it occasionally shows up in water handling systems in the form of dense black deposits. Where iron depositing bacteria exist, some manganese frequently is deposited.

Silica is found in varying concentrations in most water. Water with high silica content usually has a high alkalinity and low to medium hardness. Silica scale problems are generally confined to areas where water supplies often have high silica content. Silica is found, very weakly ionized, in solution and also as a nonionic colloidal particulate very difficult to remove. The dissolved form is referred to as "reactive " and the colloidial form as "nonreactive."

Gas Impurities

In addition to dissolved solids, natural waters contain dissolved gases. Some of the more common gases and their effects on water are described below.

Carbon dioxide is soluble in water and combines with water to form carbonic acid. In ground water, carbon dioxide accelerates the solvent effect of the water and contributes to hardness by dissolving limestone and other rocks.

The amount of dissolved oxygen in water is directly related to its temperature and pressure. Dissolved oxygen in water acts as a cathodic depolarizer, promoting corrosion.

Some well water contains hydrogen sulfide that can be released by aeration; the degree of release attainable depends on the pH of the water. More complete removal is attained at lower pH levels than at higher pH levels. Water containing hydrogen sulfide is commonly called sulfur water and has the odor of rotten eggs.

Biological Impurities

All water contains some biological activity. When organisms are present in great numbers, the chemical composition of the water may be so altered as to necessitate radical changes in methods of operation or treatment procedures. Organisms frequently cause fouling of heat transfer surfaces and clogged filters.

The most common of such organisms are the various kinds of algae, bacteria, and fungi. Microbes can be placed into functional groups as follows: acid producers that produce sulfuric acid or organic acids; slime formers that foul equipment; sulfate reducers that generate sulfides and depolarize cathodic sites on metal surfaces by consuming hydrogen; mold growers that are primarily fungi; metal ion concentrators—oxidizers that form thick bulky deposits which cover corrosive concentration cells; hydrocarbon feeders that destroy organic coatings and linings A final category is toxic microbes, such as the "legionella" bacteria which grow rapidly in water having a temperature of from 80°F to 130°F. These are deadly to the extent that a mist emanating from a cooling tower can fatally infect someone who might inhale some of the mist. The classic case of this situation occurred in Philadelphia at a legionnaires convention some fifteen years ago. The mist from an HVAC tower was carried into the ventilation system, infecting many people, a number of whom died.

The difficulties liable to be encountered by different types of water impurities are indicated in Table 1-1, Chapter 1.

MEASUREMENT OF WATER QUALITY

There are a number of terms commonly used for describing the quality of water, and there are typical units for expressing these terms. This section discusses the standard units and measurement of the following terms:

- Units of measurement
- pH
- Alkalinity
- Hardness
- Conductance and Resistance
- Total Dissolved Solids (TDS)

Parts Per Million

In reporting results of an analysis where a material is reduced to its constitutents, it is typical to express the quantity of each constituent per 100 parts of material. This gives values in "percent." For example, in analysis of an alloy, if the chemist reports a 60 percent copper content, it would mean that for each 100 parts of alloy, 60 parts of copper are present.

Reporting by percent is awkward in water analysis since the amount of materials present is usually extremely small. For example, if silica content of natural water is expressed in percent, values would range from 0.0001 percent to 0.01 percent. To avoid use of such small decimals, and the loss of significant figures, the term "parts per million" (ppm) is usually used in water analysis. Thus the above percentage range would be 1 ppm to 100 ppm.

One part per million equals one part of a constituent in one million parts of solution. As an example if a water sample contains one ppm silica there would be one ounce of silica in 1,000,000 ounces of water. Expressing results in terms of parts per million is the simplest method and allows results to be reported in whole numbers.

Grains Per Gallon

A secondary method for expressing results is in grains per gallon. The significance of this expression is less obvious than parts per million. There is a simple relationship, however, between the weight in grains of the substance being reported and the weight of a gallon of water, since one pound equals 7,000 grains.

For conversion purposes, one grain per gallon equals 17.1 parts per million.

Equivalents Per Million

A third method for expressing results of water analyses is equivalents per million. This method is often preferred by water technologists since it simplifies calculations. Results in equivalents per million are obtained by dividing the concentration in parts per million by the chemical combining weight or equivalent weight for each ion or substance. Equivalent weights can be obtained from suitable tables. For example, to convert 25 ppm of chloride to epm:

$$\frac{25\ ppm}{\text{equivalent mole atomic weight of chloride}} = \frac{25}{35.5} = 0.71\ epm$$

pH

pH is a measure of the hydrogen ion concentration, or the driving force or intensity of an acidic or basic material. It is defined as:

$$pH = \log \frac{1}{(H+)}$$

It is measured on a 0 to 14 scale, with 7 as being the neutral point Adding acid to water lowers the pH; addition of alkaline material increases the pH.

Because the pH scale is logarithmic, each decrease by one unit indicates an increase in hydrogen ions (or acidity) by a factor of ten. Similarly, an increase of one full pH unit indicates a ten-fold increase in alkalinity.

An understanding of pH is important in water treatment. In general pH ranges below 7 are considered to be increasingly corrosive; pH ranges above 7 are more conducive to scale formation.

Figures 3-3 and 3-4 present pH values of various solutions.

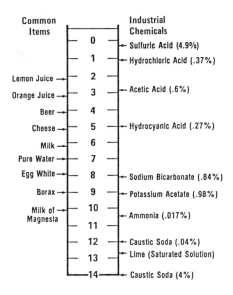

Figure 3-3. pH Values of Various Materials

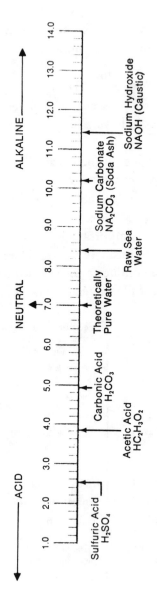

Figure 3-4. Approximate pH Values of Common Acids and Bases (100 ppm of common acids and bases dissolved in pure water)

Source: Wright Chemical Corporation

ALKALINITY

pH and alkalinity are frequently confused terms. pH is a measure of the intensity of the acid or alkaline nature of a solution. Alkalinity, on the other hand, is a measure of the quantity of anions present in a solution, generally measured in parts per million (ppm). Stated another way, it is the presence of alkaline (acid neutralizing) minerals in water. By definition, total alkalinity, also called "M Alkalinity," is that which produces a pH above the methyl orange (M) endpoint of about 4.2–4.4, and which reacts with mineral acids to produce a neutral salt at the M endpoint. Another significant alkalinity measure is "P Alkalinity" which exists over the phenolphthalein endpoint range of pH 8.2–8.4. By definition, alkalinity does not exist below pH of 4.2–4.4, and carbon dioxide does not exist above pH 8.2–8.4.

In general, the alkalinity of most natural water supplies is caused by dissolved bicarbonate salts. Rain, picking up CO_2 from the atmosphere and from the respiration of soil organisms, dissolves magnesium and calcium from a common mineral, dolomite, to produce hardness and alkalinity in ground water. Depending on the type of dissolved solids in water, it is possible for some samples to have a pH of 7.5, but with alkalinities ranging from 50 to 250 ppm. This variation in alkalinity occurs because dissolved solids have different levels of chemical activity.

Theoretically, a pH of 7.0 represents a neutral system, with higher pH being alkaline and lower pH acid. Obviously, in water chemistry the expression "neutral pH" has little or no meaning, since even at a pH of 5, alkalinity is present. To properly define a water condition, pH should be expressed quite accurately. There remains a balance between excess CO_2 and bicarbonate ions, which is measured by pH value, as shown by Figure 3-5.

If other ions are present, they can be converted to calcium carbonate equivalents by use of Table 3-2, or by the following example:

Both CO_2 and Alkalinity, as $CaCO_3$
CO_2 as CO_2 × 1.14 = CO_2, as $CaCO_3$

Example: Raw water pH = 6.9
Ratio (from Figure 3-5) = .2
Raw water titration shows:
"M" alkalinity = 150 ppm
By calculation, CO_2 (as $CaCO_3$) = 150 ppm × .2 = 30 ppm

In the region above pH 8.2–8.4, the "P" endpoint, it is convenient to use a "shorthand" device to easily calculate the balance between HCO_3, CO_3, and OH based on P and M Alkalinities. These expressions are as follows:

Below pH 9.8–10.0 (where P Alkalinity is less than 1/2 M Alkalinity):

CO_3 = 2 × P Alkalinity
HCO_3 = M − CO_3 = M −- 2P
OH   = Zero

Above pH 9.8–10.0 (where P Alkalinity exceeds 1/2 M Alkalinity):

CO_3 = 2 (M − P)
OH   = 2P − M
HCO_3 = Zero

These values are only approximate, however, since the results may be inaccurate due to interference from, for instance, ammonia, which is common in waste water.

When the pH exceeds 8.2–8.4, the free CO_2 disappears and the bicarbonate begins to convert to the carbonate ion. This progresses to a pH of about 9.8–10.0 at which point all of the CO_2 originating alkalinity is essentially in the form of carbonate.

These calculations are explained more conveniently by the table of Alkalinity Relationships, Table 3-3.

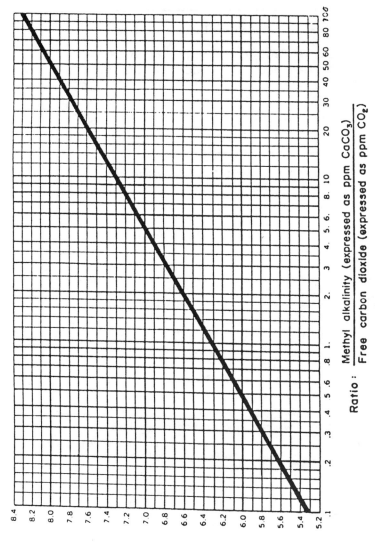

pH Values

Ratio : $\dfrac{\text{Methyl alkalinity (expressed as ppm } CaCO_3)}{\text{Free carbon dioxide (expressed as ppm } CO_2)}$

This figure shows the relationship between alkalinity, carbon dioxide content, and pH. Carbon dioxide is a factor because it forms carbonic acid when dissolved in water, which affects the pH. Note that alkalinity is expressed as parts per million of calcium carbonate, which is the reference standard.

Figure 3-5. Relationship Between Alkalinity, Carbon Dioxide and pH

Table 3-2. Conversion Factors for Calculating Alkalinity

Ion		Factor		Ion		Factor	
Ca	X	2.5	= $CaCO_3$	NO_3	X	.81	= $CaCO_3$
Mg	X	4.1	= $CaCO_3$	HCO_3	X	.82	= $CaCO_3$
Na	X	2.18	= $CaCO_3$	PO_4	X	1.56	= $CaCO_3$
K	X	1.28	= $CaCO_3$	• CO_3	X	.84	= $CaCO_3$
SO_4	X	1.04	= $CaCO_3$	• CO_2	X	1.14	= $CaCO_3$
Cl	X	1.41	= $CaCO_3$	• SiO_2	X	.83	= $CaCO_3$

• These factors based on sorption as Monovalent Ions.
  For true $CaCO_3$ equivalents, multiply factors by two.

Table 3-3. Alkalinity Relationships

Result of Titration	Hydroxide Alkalinity as $CaCO_3$	Carbonate Alkalinity as $CaCO_3$	Bicarbonate Alkalinity as $CaCO_3$
P = 0	0	0	M
P < 1/2M	0	2P	M−2P
P = 1/2M	0	2P	0
P > 1/2M	2P−M	2(M−P)	0
P = M	M	0	0

*Key: P — phenolphthalein alkalinity   M — total alkalinity.

By inspection of water analysis, the presence or absence of the following constituents and the amounts of those present may be determined as follows:

Calcium alkalinity	=	Calcium hardness or alkalinity, whichever is smaller (if equal, either one)
Magnesium alkalinity	=	Magnesium hardness, if alkalinity is equal to or greater than total hardness
Magnesium alkalinity	=	Alkalinity less calcium hardness, if alkalinity is less than total hardness
Sodium alkalinity	=	Alkalinity less total hardness
Calcium noncarbonate hardness	=	Calcium hardness less calcium alkalinity
Magnesium noncarbonate hardness	=	Magnesium hardness less magnesium alkalinity
Total noncarbonate hardness . . .	=	Total hardness less alkalinity

NOTE: In the above, if any computation yields a zero or negative result, none of that substance is present.

Another consideration is the buffer effect; this is a characteristic of certain chemicals which tend to stabilize the pH of a solution, preventing any large change when moderate amounts of acids or alkalies are added. This can be a critical factor where careful control of pH may be necessary in treating either a raw water supply or a wastewater.

As an example of this, compare the neutralization of NaOH to the neutralization of Na_2CO_3, as shown in Figure 3-6. It is obvious that it is virtually impossible to control the pH of the NaOH neutralization at pH of 7, since there is a vertical drop between pH of 9 and pH of 4.3. As a corollary to this, it is considerably easier to control the pH of an acidic waste with Na_2CO_3 than with NaOH, providing the CO_2 generation can be controlled without excessive foaming.

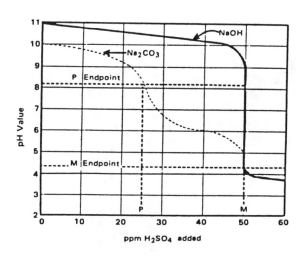

**Figure 3-6. Neutralization of Alkaline Water with H_2SO_4
(assume 50 ppm M Alkalinity)**

HARDNESS

Water hardness is caused by dissolved calcium and magnesium and occurs in significant amounts in over 85 percent of the United States and Canada. Figure 3-7 shows the variation in water hardness in the contiguous 48 states.

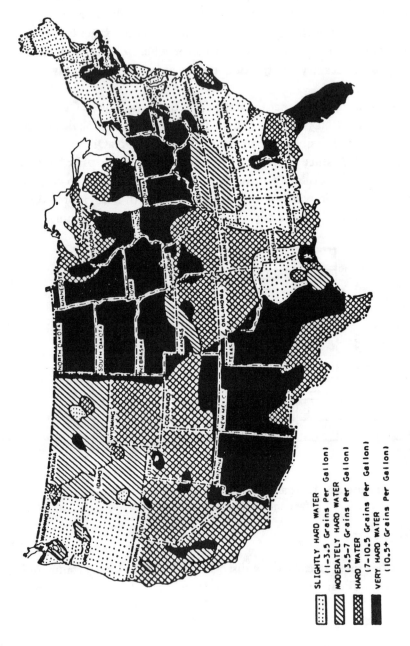

SLIGHTLY HARD WATER
(1-3.5 Grains Per Gallon)

MODERATELY HARD WATER
(3.5-7 Grains Per Gallon)

HARD WATER
(7-10.5 Grains Per Gallon)

VERY HARD WATER
(10.5+ Grains Per Gallon)

Figure 3-7. Water Hardness Throughout the Contiguous United States

Source: U.S. Geological Survey

The primary source of water hardness is common limestone (calcium carbonate). Other sources include calcium sulfate, magnesium sulfate, and either calcium or magnesium chloride. Water hardness is the chief cause of scale buildup in heat exchange equipment and must be reduced by treatment.

Hardness is expressed as parts per million or grains per gallon of calcium carbonate equivalent and is classified as follows:

Classification	Grains Per Gallon	Parts Per Million
Soft or Slightly Hard	0– 3.5	0– 60
Moderately Hard	3.5– 7.0	61–120
Hard	7.0–10.5	121–180
Very Hard	more than 10.5	More than 180

Hardness is further defined as temporary, or permanent, or some combination of the two categories based upon the presence and ratio between carbonates and sulfates. Water hardness in the U.S. is defined as one degree of hardness equals one grain gallon which equals 0.017 gm/liter of calcium carbonate or the gram molecular weight adjusted equivalent. The scientific (French) system defines one degree equal to one part calcium carbonate (or equivalent) per 100,000 parts water, or 01.01 gm/liter. In practice, soft water is around 5 degrees or less and very hard water is above 30 degrees.

Hardness Remediation

Temporary hardness is due to bicarbonates and the designation "temporary" is used because simply boiling the water will precipitate the normal carbonates formed by the rather easy decomposition of bicarbonate ions.

Permanent hardness is not eliminated so easily because the sulfates of alkali earth and alkali metals in most water supplies are quite soluble, so they are not precipitated by chemical changes produced by boiling. Treatment of permanent hardness requires the use of chemicals that leave residues in the processed water. Treatment of temporary hardness can be done using chemicals that leave little residue and the most favored mineral chemical for this purpose is ordinary lime. Accordingly, when dissolved solids of temporary hardness are present, the adding of a relatively cheap mineral will cause a (desirable) reduction in process water mineral content.

The use of lime to correct hardness seems contrary to the general perception that lime causes hardness. The misconception is understandable because lime, or calcium oxide, is derived from the same mineral carbonate responsible for temporary hardness. Therefore, instead of making things worse, slaked lime or calcium hydroxide provides a strongly ionized base to reduce the soluble acid carbonate ions to precipitates of the more insoluble carbonates or hydroxides.

The following rules for chemical softening apply:

(1) Rain water + carbon dioxide leaches carbonates,
$$H_2O + CO_3 \leftrightarrow H_2CO_3 \text{ (carbonic acid)}$$
$$CaCO_3 + H_2CO_3 \rightarrow 2Ca(HCO_3)_2 \text{ (acid carbonate)}$$
forming bicarbonates of greater solubility. These contribute to the temporary hardness of water.

(2) Permanent hardness minerals, Ca and Mg sulfates, are not precipitated by boiling.

(3) Ordinary liming of water precipitates the carbonates. A slight excess of lime allows Mg to precipitate as the far more insoluble hydroxide (0.02 g/liter).

(4) Permanent hardness can be precipitated easily by crude soda ash, but leaves a residue of sodium sulfate in solution.
$$CaSO_4 + Na_2CO_3 \text{ - - - } Na_2SO_4 + CaCO_3 \downarrow$$

(5) Sodium hydroxide (or phosphate) will handle both forms, but can be expensive for large water volumes.

CONDUCTANCE AND RESISTANCE

Water quality is also expressed in terms of conductance (conductivity) and resistance (resistivity). Absolutely pure water contains no ions and therefore has no means to transport electric energy. When ions are present as impurities, the resulting solution has the ability to conduct electricity, in proportion to the number and type of ions. The ionizible solids in the water are termed "electrolyte" and include the total ionizible solids of the water. It is distinguished from nonionic solids, such as organic matter or bacteria. Demineralizers remove electrolytes, not the nonionic solids and bacteria. It follows that resistance increases as the impurity level decreases. Electric resistance is defined by Ohm's Law as:

$$\text{Resistance} = \frac{\text{Voltage}}{\text{Current}}$$

Current is expressed in amperes, potential in volts, and resistance in ohms. Conductance is the inverse, or reciprocal, of resistance, and is assigned the unit "mho," which is the inverse of "ohm." Inasmuch as the area of the measuring electrodes will affect the measurement, conductance has been standardized as the current which will travel across a one-centimeter cube. Because conductance (expressed as mhos in water) is usually a small number, "micromhos" or one-millionth of a mho (designated "umho/cm") is the standard parameter. An increase of ion concentration in the water decreases the resistance, and increases the conductance. Figure 3-8 shows a relationship between dissolved solids, resistance, and conductance. Lacking information regarding the type of ions in the water, the common assumption is made that conductance X 0.5 = total dissolved solids (TDS). This multiplier is variable depending on source water conditions. The multiplier may be determined, if necessary, by measuring the conductance, then evaporating a fixed quantity of water and measuring the solids remaining. The TDS, then, is the relationship between the weight of the residual solids and the weight of the evaporated water.

$$\frac{1,000,000}{OHMS} = MICROMHOS \qquad \frac{1,000,000}{MICROMHOS} = OHMS$$

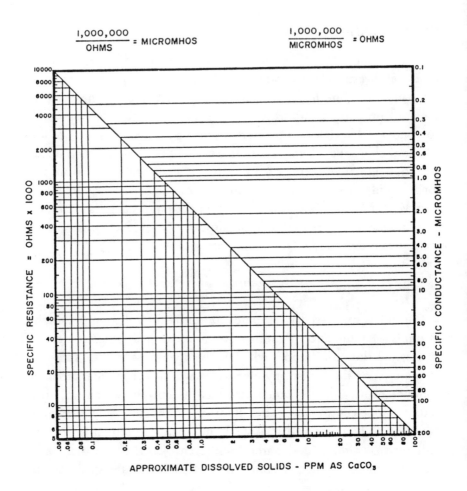

APPROXIMATE DISSOLVED SOLIDS - PPM AS CaCO₃

**Figure 3-8. Conversion for OHMS and MICROMHOS
with Approximate Dissolved Solids**

Source: Culligan International Company

The various ions have different conductances. Temperature also affects the conductance so there is no fixed rule for determining conductance short of making measurements. With a mixed bed, or with a two-bed weakly basic demineralizer, average water will have a conductance of 2.5 micromhos for each ppm of electrolyte, assuming the electrolyte is Sodium Chloride (NaCl) expressed as Calcium Carbonate ($CaCO_3$) at a temperature of 25°C and free of gases such as Carbon Dioxide (CO_2) or Ammonia (NH_3). If gases are present they affect the resistance. They must either be removed, or determined and corrected. For two-bed demineralizers with strongly basic anion resin the relationship is 5 micromhos for each ppm of electrolyte. This is better shown by Table 3-4. The table shows typical values for conductivity of water passed through different kinds of demineralizers.

Table 3-5 shows how electrical resistance varies with water purified by several techniques.

TOTAL DISSOLVED SOLIDS (TDS)

All water contains dissolved solids that have been collected from the atmosphere, earth, and oceans. The classification of water by concentration of dissolved solids is as follows:

Description of Waters	Concentration of Dissolved Solids (ppm)
Fresh	0– 1,000
Brackish	1,000– 10,000
Salty	10,000–100,000
Sea Water	33,000– 36,000
Brine*	100,000+

*As found in the Dead Sea, Great Salt Lake, and brine wells

When water is evaporated from a container, dissolved solids are deposited as "scale." When water is used in industry, it generally is not boiled to dryness but is subjected to some heating or cooling. Under these conditions scale forms from the "unstable" salts of calcium and magnesium present. Scale-forming compounds are known collectively as "water hardness." They deposit since there is a tendency to exceed their solubility limits when the temperature

Table 3-4

Resistance (in ohms)	Conductivity in microhms	Electrolyte (NaCl) in ppm as CaCO₃ (Mixed Bed or Two Bed Weakly Basic Resin)	Electrolyte in ppm as CaCO₃ (Two Bed with Strongly Basic Resin)
10,000,000	0.33	0.13	0.06
2,500,000	0.40	0.16	0.08
2,000,000	0.50	0.20	0.10
1,000,000	1.00	0.40	0.20
500,000	2.00	0.80	0.40
100,000	10.00	4.00	2.00
50,000	20.00	8.00	4.00

Table 3-5. Resistance of Water Obtained from Various Sources

Type of Water	Resistance, ohms/cm
Theoretical maximum quality (calculated)	26,000,000
Water after 28 distillations in quartz	23,000,000
Water treated in a Monobed resin system (strongly acidic — strongly basic system)	18,000,000
Water after three distillations in quartz	2,000,000
Water after three distillations in glass	1,000,000
Water in equilibrium with the carbon dioxide in the atmosphere	700,000
Water after a single distillation in glass	500,000
Approximate quality of U.S.P. distilled water	100,000–500,000

increases. In addition, the presence of other substances in a water system sometimes modifies the rate of scale formation.

To illustrate this point, consider the solubilities of some dissolved solids commonly found in water (Table 3-6).

Also notice Figure 3-9, which indicates the methods of treatment with respect to Total Dissolved Solids.

Water Quality

Table 3-6. Solubility in Water (ppm as calcium carbonate)

| | Water Temperature: | |
	32 Degrees F	212 Degrees F
Calcium		
Bicarbonate	1,620	Decomposes to carbonate
Carbonate	15	13
Magnesium		
Bicarbonate	37,000	Decomposes to carbonate
Carbonate	101	75
Sodium		
Bicarbonate	38,700	Decomposes to carbonate
Carbonate	61,400	290,000

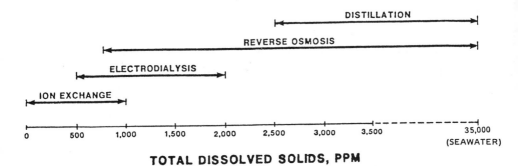

TOTAL DISSOLVED SOLIDS, PPM

Figure 3-9. Effect of TDS on Methods of Treatment

When water is heated, bicarbonates of calcium, magnesium, and sodium decompose to carbonates. In the case of calcium and magnesium this lowers the solubility and they tend to deposit as scale. With sodium, however, the solubility is increased and scaling is not a problem.

The problem associated with calcium and magnesium salts in scale formation is illustrated by an example. A boiler producing 100,000 pounds of steam per hour will use 300,000 gallons of water a day. If untreated Lake Michigan water with only moderate hardness is used, about 340 pounds of calcium and magnesium salts will enter the boiler each day. On an annual basis this amounts to over 100,000 pounds of potential scale in one year.

WHAT IS "GOOD QUALITY" WATER?

The requirement for water quality is determined by the end use of the water. For example, Table 3-7 shows the results of three water analyses in comparison with the U.S. Navy specification for shore-to-ship steam quality. For selected cooling applications, Sample A (sea water) may be of sufficient quality. Sample B, taken from a municipal water supply, may be suitable for heat exchangers and human consumption. Sample C, taken from a distribution main at a major naval facility, is of sufficient quality for shoreside process steam use and base heating. Sample C, however falls far short of purity requirements specified by the Navy for delivery to ships. The Navy specification is necessarily strict because condensate is collected by the shipboard boiler system. Most shipboard boilers operate at high pressure, and high quality water is a mandatory requirement.

THE LANGELIER INDEX

One important measure of corrosivity of water is a factor known as the calcium carbonate saturation index. If the index is a positive

Table 3-7. Comparison of Water Quality

| | Water Samples | | | |
	A	B	C	D
	Sea Water	Municipal water supplies for Navy base (1982)	Condensate from shore steam headers at Navy base supplying steam to ships (1982)	U.S. Navy requirement for shore to ship steam purity
pH	7.5–8.4	7.2–8.7	11.0–11.1	8.0–9.5
Conductivity Micromhos	12,000	40–300	885–890	25 Max
Dissolved Silica (ppm)	10	2.3–11.1	.4–1.4	.2 Max
Hardness (ppm)	6,250	15–132	2	.1 Max
Total Dissolved Solids (ppm)	34,450	35–205	619–710	No requirement

number, the water has a tendency to precipitate calcium carbonate, causing scale; whereas, a negative index corresponds to water that is likely to be corrosive. W. F. Langelier developed a procedure for calculating the calcium carbonate saturation index of a water. (Figure 3-10)

Several parameters are needed for the calculation:

- Total alkalinity (methyl orange)
- Calcium hardness
- Total dissolved solids
- pH
- Temperature

For potable water, a close approximation of the index can be obtained by the use of charts (Figure 3-10), a special slide rule, or a

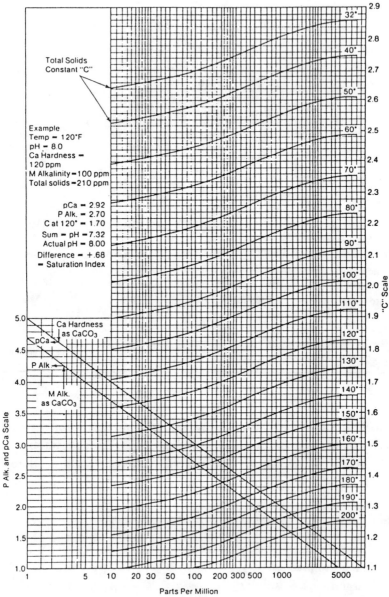

The stability index is empirical but has been related to actual experience through the Ryznar Stability Index.

Figure 3-10. Langelier Saturation Index

simple formula and tabulated data. In many cases, there is insufficient information to indicate whether a scale will form and whether it will be dense, uniform, or protective. The flow characteristics of a system tend to influence both the amount of scale that develops and the uniformity of the deposit. In practice, indications are that scale will start to deposit at slightly supersaturated conditions, depending on pH.

A high positive number for the Langelier Index indicates that heavy deposits can be expected. These can lead to increased frictional resistance, clogging of valves and controls, and decreased heat transfer rates in water heaters and boilers.

Realistically, the Langelier Index is not a reliable indicator of whether copper alloys will be subject to corrosion. In fact, the Langelier Index indicates that distilled or deionized water is corrosive. A negative index indicates that the metal in the distribution system must rely on itself for protection against corrosion. When the oxygen in very pure water contacts metal, corrosion will occur because there are no reactive solids to form a passive coating on the metal. This explains the corrosivity of dimeneralized water.

Other factors come into play. Analyses of many waters high in dissolved carbon dioxide show acceptable Langelier indices, but pitting corrosion takes place nevertheless. In addition, water that has undesirable properties at room temperature (an unacceptable Langelier Index) will often become innocuous when heated.

It is evident that the Langelier Index should be used carefully and on a system-by-system basis for evaluating corrosion potential. An empirical method for predicting scaling tendencies of water, based on a study of operating results with water of various saturation indices, has been developed and is called the Ryznar Stability Index. This index, which is calculated by subtracting the pH from the doubled Langelier Index, is often used in combination with the Langelier Index to improve accuracy in predicting scaling or corrosion tendencies. The following chart (Figure 3-11) illustrates how to use the Ryznar Stability Index.

Ryznar Stability Index

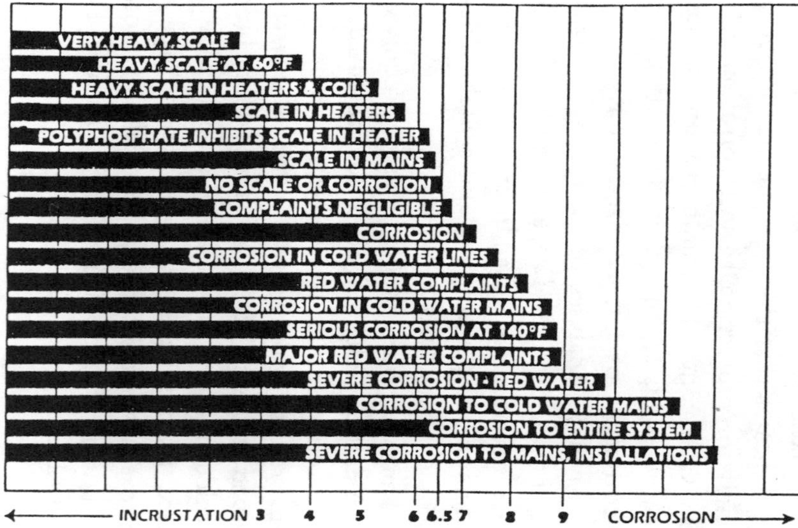

An empirical method for predicting scaling tendencies of water based on a study of operating results with water of various saturation indices.

where: Stability Index = 2pHs-pH
        pHs = Langelier's Saturation pH

Ryznar Stability Index	Tendency of Water
4.0 - 5.0	heavy scale
5.0 - 6.0	light scale
6.0 - 7.0	little scale or corrosion
7.0 - 7.5	corrosion significant
7.5 - 9.0	heavy corrosion
9.0 and higher	corrosion intolerable

STABILITY INDEX

VERY, HEAVY, SCALE
HEAVY SCALE AT 60°F
HEAVY SCALE IN HEATERS & COILS
SCALE IN HEATERS
POLYPHOSPHATE INHIBITS SCALE IN HEATER
SCALE IN MAINS
NO SCALE OR CORROSION
COMPLAINTS NEGLIGIBLE
CORROSION
CORROSION IN COLD WATER LINES
RED WATER COMPLAINTS
CORROSION IN COLD WATER MAINS
SERIOUS CORROSION AT 140°F
MAJOR RED WATER COMPLAINTS
SEVERE CORROSION - RED WATER
CORROSION TO COLD WATER MAINS
CORROSION TO ENTIRE SYSTEM
SEVERE CORROSION TO MAINS, INSTALLATIONS

INCRUSTATION 3   4   5   6 6.5 7   8   9   CORROSION

Figure 3-11. Ryznar Stability Index

Notice that when your index is about 6.5–7.0, problems and complaints practically disappear. But go below 6.5 and you can prepare yourself for scale and liming. Go above 7.0 and corrosion's got the best of you . . . red water complaints increase and pipe and main life expectancy decreases.

A low Ryznar Index, left untreated, means calcium and magnesium salts precipitate, leaving a build-up of mineral deposits on your pipes. Flow decreases—complaints increase: low pressure, rings around tubs and basins, difficulty in rinsing off soap and short water-heater lives are just a few of the things you can blame on hard water.

A high Ryznar Index, left untreated, will result in equally unstable water.

4

Introduction to
Water Treatment

The type of water treatment required is determined by the quality of the raw water and the end use of the treated product. In this chapter, various methods of water treatment are described, including circumstances governing the use of each treatment.

An overview of water treatment methods appears in Figures 4-1 and 4-2. Figure 4-1 provides a general indication of end uses for various qualities of water. It is important to note that water treatment depends on constituents present. Therefore, Figure 4-1 cannot be used solely as a guide to treatment.

The intended use of water determines the extent of treatment. For example, Mississippi River water used for some cooling applications may require only rough screening. Clarification and filtration may be required for other uses.

Table 4-1 shows typical applications for various treatment processes. Table 4-2 displays a comparison of treatment processes as a function of impurities to be removed.

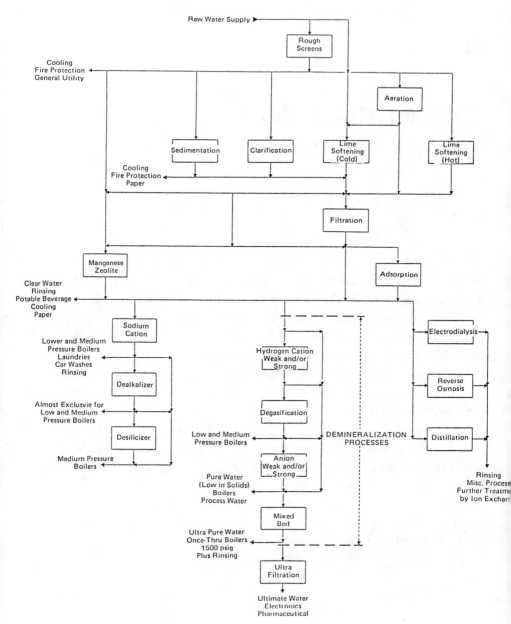

Figure 4-1. Water Trestment Processes and Typical Water Uses

Courtesy of Drew Industrial Division, Ashland Chemical, Inc. Subsidiary of Ashland Oil, Inc.

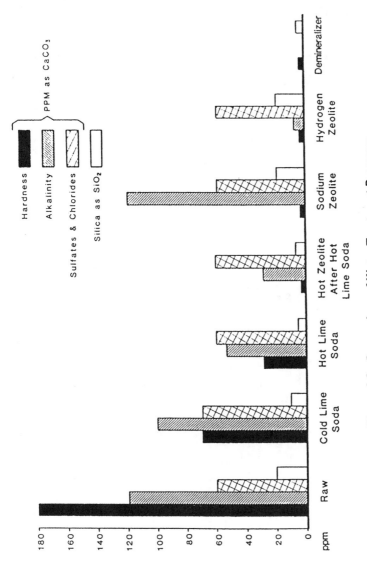

Figure 4-2. Comparison of Water Treatment Processes

Courtesy of Drew Industrial Division, Ashland Chemical, Inc. Subsidiary of Ashland Oil, Inc.

Table 4-1. Typical Raw Water Analyses and Operating Results[1]
(mg/1, except for pH)

Water Constituents (ppm)	Expressed As	Raw Water	After Clarification and Filtration	After Cold Lime Softening and Filtration	Hot Lime	After Clarification, Filtration, and Sodium-Cation Exchange Softening	After Clarification, Filtration, and Demineralization[2]	Hydrogen Cycle Ion Exchange[2]	Reverse Osmosis[3]	Electrodialysis[4]	Distillation[5]
CATIONS											
Calcium	$CaCO_3$	51.5	51.5	38.7	17	1.0	0	0	6	17	0
Magnesium	$CaCO_3$	19.5	19.5	17.5	7.2	1.0	0	0	2	7	0
Sodium	$CaCO_3$	18.6	18.6	18.6	18.6	87.6	1-2	0.1	7	13	1-2
Potassium	$CaCO_3$	2	1.8	1.8	2	1.8	0	0	1	2	0
		91.6	91.4	76.6	44.8	91.4	1-2	0.1	16	39	1-2
ANIONS											
Bicarbonate	$CaCO_3$	56.8	47.8	0	0	47.8	0	0	9	14	1-2
Carbonate	$CaCO_3$	0	0	33.0	0	0	0	0	0	0	0
Hydroxide	$CaCO_3$	0	0	0	0	0	1-2	0.1	0	0	0
Sulfate	$CaCO_3$	21.8	30.8	30.8	31.8	30.8	0	0	2	13	0
Chloride	$CaCO_3$	12	12.0	12.0	12	12.0	0	0	5	7	0
Nitrate	$CaCO_3$	1	0.8	0.8	1	0.8	0	0	0	0	0
		91.6	91.4	76.6	44.8	91.4	1-2	0.1	16	39	1-2
OTHER											
Iron	Fe	0.16	Nil	Nil	Nil	Nil	Nil	0	0	0	0
Silica	SiO_2	9	9.0	9.0	1	Nil	Nil	0.01	5	5	1
pH	units	7.1	6.0-8.0	9-11	9-11	6-8	7-9	6.5-7.5	7.0-9.0	6.5-7.5	5.0-6.0

[1] Provided only to show approximate effect on water quality given. Higher or lower dissolved constituents would be affected differently (efficiencies).
[2] Mixed-bed demineralizer.
[3] Based on 2 stage: % of salt passage based on actual pilot test of 500 ppm water.
[4] Reverse polarity, 2 stage ED.
[5] Multistage Flash unit. This method is generally not cost effective for potable waters, only seawater. However, it is included for comparative purposes.

Table 4-2. Summary of Suggested Treatment Methods versus Impurities

Ionic Impurities

Impurities	Methods
Cations 1. Calcium and magnesium	a. Cold, warm, or hot lime-soda process precipitation, settling and filtration
	b. Ion exchange
2. Sodium, potassium and ammonium	a. Hydrogen cation exchange, if bicarbonate present exceeds total hardness
	b. Demineralization
3. Iron and manganese	a. Oxidation (aeration) and precipitation, settling (if high amounts present), and filtration (chlorine and alkali may be needed)
	b. Filtration through manganese zeolite
	c. Ion exchange
Anions 4. Alkalinity	a. Lime process as in 1a, but without soda ash
	b. Hydrogen cation exchange
	c. Chloride anion exchange salt-splitting (dealkalization)
5 Sulfate, chloride, nitrate and phosphate	a. Demineralization
6. Silica	a. Absorption by ferric hydroxide precipitated by adding ferric sulfate: settling and filtration follow
	b. Absorption by magnesium hydroxide, formed when lime or dolomitic lime is added: settling and filtration follow: adding activated magnesia with lime in warm or hot process is helpful.
	c. Hydroxide anion exchange salt splitting (desilicization)
	d. Demineralization

Nonionic Impurities

Impurities	Methods
1. Turbidity and suspended matter	a. Filtration alone for small amounts of turbidity, adding coagulant directly ahead of filters if clearer effluent desired
	b. Coagulation, settling and filtration for larger amounts of turbidity: prechlorination usually beneficial; alkali addition, if needed for optimal pH value; coagulant aid often improves the floc
2. Color	a. Same as 1b, but addition of clay or other weighting agents, to densify floc; if water has low amounts of suspended matter
3. Organic matter	a. Same as 1b b. Addition of oxidizing agents, such as chlorine or permanganate c. Absorption by powdered or granular activated carbon d. Absorption by anion exchangers
4. Colloidal silica	a. Same as 1b b. Recirculation of boiler blowoff through demineralizer

Table 4-2. Summary of Suggested Treatment Methods versus Impurities (Continued)

Nonionic Impurities

Impurities	Methods
5. Plankton and bacteria	a. Same as 1b b. Superchlorination
6. Oil	a. Same as 1b b. Addition of preformed alum floc and filtration
7. Corrosion products in condensate	a. Filtration with cellulose filter and b. Cation exchanger c. Ammoniated cation exchanger for heater drains. d. Combined filtration and ion exchange with mixed bed demineralizer

Gaseous Impurities

Impurities	Methods
1. Carbon dioxide	a. Aeration: open aerator b. Aeration: degasifier (decarbonator) or forced-draft aerator c. Vacuum deaerator d. Heater deaerator for boiler feed
2. Hydrogen sulfide	a. Aeration as in 1a or 1b b. Chlorination c. Aeration plus chlorination
3. Ammonia	a. Hydrogen cation exchange, if the ammonia is present as ionic NH_4^+
4. Methane	a. Aeration as in 1a or 1b
5. Oxygen	a. Vacuum deaerator b. Heater deaerator for boiler feed c. Addition of sodium sulfite or hydrazine d. Anion exchanger regenerated with sodium sulfite, hydrosulfite, and hydroxide
6. Excess residual chlorine	a. Dechlorination by addition of reducing agents such as sodium sulfite, hydrazine or sulfurous acid b. Absorption by powdered or granular activated carbon c. Filtration through granular calcium sulfite

Courtesy of Drew Industrial Division, Ashland Chemical, Inc. Subsidiary of Ashland Oil, Inc.

PRELIMINARY TREATMENT

Raw water is generally pretreated by passing the water through a screen to remove the larger particles of suspended matter. Preliminary treatment for potable water also can consist of:

- Aeration, to remove undesirable gases
- Clarification, to remove smaller particles of suspended matter
- Filtration, to remove minute suspended particulates or to supplement clarification

Aeration

Aeration is a mechanical process that intimately mixes water and air. Mixing may be accomplished in the form of a thin film, drops, or spray. Aeration is based on establishment of a state of equilibrium between gases in the water and gases in the atmosphere.

Aeration is used to release undesirable gases from water. It also is used to remove corrosive gases, lower chemical costs in water treatment processes, remove undesirable metals, and eliminate odor, tastes, and colors from biologically contaminated waters. It is also used to oxidize dissolved iron and convert it to filterable particles.

The removal of gases by aeration is accelerated by an increase in temperature, increase in aeration time, increase in the volume of air in contact with the water, and increase in the surface area of water exposed to the air. The efficiency of aeration is greater when the concentration of the gas to be removed is high in the water and low in the atmosphere. For example, aeration efficiency is higher with a water containing 100 ppm carbon dioxide, than with one containing only 10 ppm.

The most common method of aeration causes water to fall through the air using devices such as spray nozzles, cascade over fill material, and forced-draft-type aerators.

The air diffusion method of aeration causes air to bubble through water. This method is usually confined to treatment of relatively low flows.

Many variations of the water-fall principle are used. The simplest is the use of vertical risers to discharge water by free fall into a basin. The risers can operate on the available head of water, and efficiency of aeration is increased by making the fall as great as practical. The addition of steps or shelves to break the fall and to spread water into thin sheets or films increases contact time and is a further refinement. Spray nozzles are another variation of this principle and can provide efficient removal of dissolved gases. Space requirements frequently limit the usefulness of water-fall aeration.

Air diffusion is accomplished by pumping air into water through perforated pipes, strainers, porous plates, or tubes. Aeration by diffusion is theoretically superior to water-fall aeration because fine bubbles rising through water are continually exposed to fresh liquid surfaces, providing maximum water surface per unit of air. Also, the velocity of bubbles ascending through water is much slower than the velocity of free-falling drops of water, providing longer contact time. For gas removal, however, air requirements by the diffusion method are higher than for odor removal. The cost of air supply is a factor restricting application of this type of aeration in industrial water conditioning.

One major application of aeration is removal of corrosive gases, such as carbon dioxide or hydrogen sulfide. On many occasions aeration is used to remove carbon dioxide liberated by a treatment process. For example, in boiler feedwater conditioning, it is common practice to acid-treat effluent from a sodium zeolite softener to reduce alkalinity of boiler feedwater. Carbon dioxide is produced as a result of acid treatment and aeration is employed to rid the water of this corrosive gas. Similarly, when influents of hydrogen and sodium zeolite units are blended, the carbon dioxide formed is eliminated by aeration.

Lime softening removes carbon dioxide before it can enter equipment. Economics favor removal of high concentrations of carbon dioxide by aeration rather than by chemical precipitation with lime.

Aeration also removes methane from certain ground waters. Since methane constitutes an explosion hazard on release in confined spaces, aeration is employed for its removal. Electric motors in the vicinity are undesirable and forced draft deaerators are not employed in this application. Iron and manganese can be removed by a variety of methods. In all but a few of the processes, aeration plays a part. Aeration is a convenient method of supplying the oxygen required for oxidation of iron and manganese. Carbon dioxide released by aeration raises the pH of the water, aiding in precipitation of oxidizing iron and manganese. Aeration alone is not considered to be an efficient method for removal of iron and manganese; it is only part of the treatment required.

Limitations on Aeration

Theoretically, at 68°F it is possible to reduce the carbon dioxide content of water to 0.5 ppm, by aeration. Practically, complete gas removal is not economical, and reduction of carbon dioxide to 10 ppm is normally a realistic limit.

Although removal of free carbon dioxide increases the pH of water and renders it less corrosive, aeration saturates the water with dissolved oxygen. When the dissolved oxygen content is already high, no undesirable effect is produced. In the case of well water, high in carbon dioxide but low in oxygen, aeration may exchange one corrosive gas for another.

The efficiency of aeration is greatest when the initial concentration of the gas to be removed is considerably above equilibrium. Therefore, when water contains only a small amount of carbon dioxide, lime treatment is usually more efficient than aeration.

Aeration alone will not completely remove hydrogen sulfide. Either pH reduction or chlorination must also be performed. A specialized case of aeration is the use of air stripping systems for the removal of volatile organic contamination from water. Air stripping systems are designed to remove trace contamination of water and groundwater by volatile organic compounds (VOCs). Hazardous substances such as benzene and trichloroethylene are slightly soluble in water at ordinary temperatures and can pollute surface and groundwater on contact.

Air strippers are an economical and efficient means of reducing trace VOC contamination to acceptable levels. Continuous, fully automatic equipment has been used to remove aromatics such as benzene, toluene, and xylene down to levels in the low parts per billion range.

Air stripping is essentially a countercurrent air/liquid extraction process whereby relatively large volumes of air are passed upwatd through a packed column. Water polluted with VOCs is pumped into the top of the column and allowed to fall by gravity through the upward rushing stream of air. The internals of the air stripper cause the water to form films and droplets with extensive surfaces exposed to the high velocity air flow. The volatile contamination then vaporizes into the moving air and is vented through the top of the column (Figure 4-3). In situations where the vented air from the top of the air stripper is objectional, demisters and/or special absorption units can be supplied as accessory equipment; this may consist of a vapor-phase carbon absorber, while the clean water is released back onto the environment or through a liquid-phase carbon absorber.

To date, the treatment techniques which have received the most attention for removing VOCs from groundwaters have been aeration and granular activated carbon (GAC) adsorption. Conventional water treatment techniques, such as flocculation, clarification and filtration, do not remove VOCs to any appreciable degree. Other treatment techniques, including reverse osmosis and adsorption using synthetic resins, have been found to be effective in removing VOCs, but have not been economically and technically proven as have aeration and carbon adsorption.

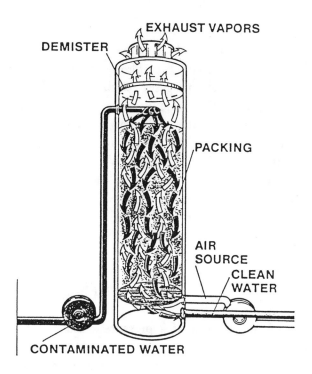

Figure 4-3. VOC Removal

Iron and Manganese Removal

Iron and manganese can be removed with a variety of methods. In all but a few of the processes used, aeration plays a part since it is a convenient method of supplying the oxygen required for oxidation of iron and manganese. Also, the carbon dioxide release by aeration raises the pH of the water, aiding in the precipitation of oxidized iron and manganese. However, aeration alone should not be considered an efficient method to remove iron and manganese, but rather only a part of the treatment required.

The description of a typical system is illustrative of the methods used to remove iron. This system employs the fundamental oxidation concept in removing iron (Fe) and manganese (Mn) from water.

Iron and manganese normally found in water are in the reduced state of Ferrous (Fe^{++}) and Manganuos (Mn^{++}) complexes which are insoluble in water. The precipitates of these metals are then filtered out. The undesirable gases (Carbon Dioxide, Hydrogen Sulfide, etc.) dissolved in this same water are concurrently released during this oxidation process due to the Partial Pressure Phenomena.

Aeration is accomplished by admitting a metered amount of compressed air into the fluid tank. The small amount of compressed air contains enough oxygen to oxidize the dissolved ferrous iron, causing it to change into ferric iron, which is insoluble in water. From the aerator, the water passes into a filter, where the precipitated material is removed. Manganese is removed from the water concurrently with the iron.

If the contamination is magnetic iron oxide, it can be readily removed by an electromagnetic filter (EMF), such as is produced by the Babcock and Wilcox Company.

The B&W EMF consists of a pressure vessel containing a matrix of ferritic stainless steel spheres that are magnetized by an electromagnetic coil. Corrosion products are efficiently removed when fluid containing magnetic particulates passes through the sphere matrix. Limited amounts of nonmagnetic materials are also removed, depending on system conditions and particulate characteristics. EMFs have achieved greater than 90 percent removal efficiency for magnetic iron oxide present in concentrations as high as 100 ppm and as low as 0.01 ppm, with particulate sizes between 0.05 and 10.0 microns. Thus, the EMF has been proven to be effective in removing particles of micron and even submicron sizes.

Figure 4-4 illustrates two types of electromagnetic filters, the sphere matrix of the B&W EMF device, and a filamentary matrix, available from Boliden-Allis. It is slightly more complex than the B&W device. The filamentary matrix filter is slightly more effective than the B&W unit in filtering out weakly magnetic material such as hematite. Experience has shown that the one factor greatly affecting the efficiency of EMFs is the percentage of magnetite or strongly

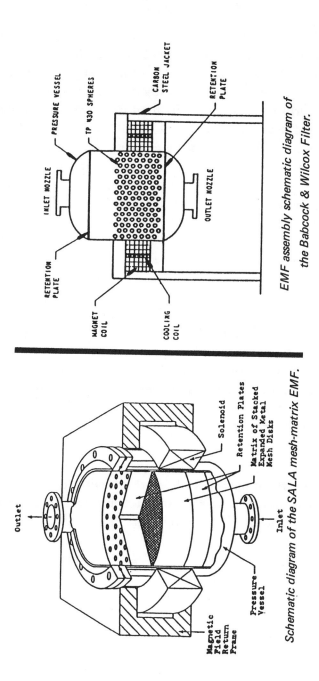

EMF assembly schematic diagram of
the Babcock & Wilcox Filter.

Schematic diagram of the SALA mesh-matrix EMF.

Figure 4-4. Electromagnetic Filter Configurations

magnetic species present in the influent. A reducing environment and high temperatures make ideal conditions for the use of an EMF. The pH of the process stream will affect the solubility of the iron in the water, and since the filter cannot remove dissolved iron, will affect filter performance relative to total iron removal. Ideally, for best results, pH should be between 9.3 and 9.5. Flowrate is also an important factor, and should not be greater than 1 foot per second through the matrix area of the sphere device, and not more than 450 gpm per square foot of media. It is known that some of the weakly magnetic ferrous materials will agglomerate with the strongly magnetic particles, and be removed along with them. Copper, chromium and nickel, on the other hand appear to form compounds with the magnetite and are thus easily removed. Experience has shown that a sphere matrix filter where over 50 percent of the iron present is magnetite, will remove 60 percent of the copper present in the stream. This proven technology is very useful for treatment of hot condensate, or boiler water.

Iron in condensate may also be removed by filtration through manganese greensand zeolite. This is a simple procedure to clean up contaminated condensate. Potassium permanganate is used for regeneration. The information displayed on the following page is issued by resin manufacturers regarding the use of this technique.

Air Flotation

Dissolved air flotation is an alternative clarification process to sedimentation in potable water treatment. Quite common in Europe, it is not widely used in the USA. A recent EPA project has provided research into the parameters for successful operation of the method. Performance depends on raw water quality, pretreatment, bubble size, and bubble volume concentration. The method is effective in the removal of humic substances, color, or algae.

To remove particles, the particles must first be destabilized with coagulants. A short flocculation period ahead of flotation may be beneficial; however, long flocculation times are not necessary.

25

CAPACITY OF MANGANESE GREENSAND ZEOLITE, IONAC M-50
Average gallons of treated water (thru-put), per cu ft of Zeolite
(based on 4 oz KMn O_4 per cu ft regeneration)

Influent Content (ppm)	For Total Iron Only Fe (gal.)	For Total of Iron Plus Manganese Fe and Mn (gal.)	For Hydrogen Sulfide Only H_2S (gal.)
0.5	24000	13800	7800
1.0	12000	6900	3900
1.5	8000	4600	2600
2.0	6000	3450	1950
3.0	4000	2300	1300
4.0	3000	1725	975
5.0	2400	1380	780

NOTE: Water should contain minimum of 4 ppm Alkalinity for every ppm of Iron or Manganese.

Suggested Operating Conditions	U.S.	Metric
Maximum Water Temperatures	100°F	37°C
Maximum Iron (Fe^+)	10 ppm	10 ppm
Maximum Iron plus Manganese (Fe^+, Mn^+)	10 ppm	10 ppm
Maximum Hydrogen Sulfide (H_2S)	6 ppm	6 ppm
pH Range	6.0-8.0	6.6-8.0
Rising Space (Freeboard)	35-45%	35-45%
Service Flow Rate: Intermittent	6 US gpm/sq ft	14.5 m/hr
Continuous	4 US gpm/sq ft	9.8 m/hr
Regenerant: (Potassium Permanganate)	KMn O_4	KMn O_4
Amount per cu ft	3-4 oz	85-113 grams
Concentration*	2%	2%
Contact Time	40 minutes	40 minutes

*Note: Water temperature directly affects the dissolving of Potassium Permanganate, with most effective range being 50°F to 72°F.

at 50°F 4 oz KMn O_4 will dissolve in (1) gallon of H_2O
at 72°F 8 oz KMn O_4 will dissolve in (1) gallon of H_2O

Water above 72°F may have adverse effect on oxidizing ability of KMn O_4.

Source: Ionac Division of Sybron Corp., Birmingham, NJ

Dissolved air flotation is an efficient process due, in part, to the small bubble size produced by the process (bubble diameters of 10 to 100 p.m.).

This EPA project is fully documented in "Removal of Humic Substances and Algae by Dissolved Air Flotation," available from National Technical Information Service (703) 487-4650.

Clarification

Suspended matter in raw water supplies is generally removed by coagulation, flocculation, and sedimentation, often referred to as conventional clarification.

Coagulation is the process of destabilization by charge neutralization. Once neutralized, particles no longer repel each other and are brought together.

Flocculation is the process of bringing together destabilized or "coagulated" particles to form a larger agglomeration or floc.

Sedimentation, or settling, is the physical removal from suspension that occurs once the particles have been coagulated and flocculated. Sedimentation (subsidence) alone, without prior coagulation, can only remove relatively coarse suspended solids.

In-line clarification is useful in raw water treatment. This process consists of filtration of water for removal of particulate matter. Either coagulant chemicals or polymers are normally added to improve operating efficiencies.

Finely divided particles (colloidal particles) suspended in normal surface water repel each other because their surfaces predominantly carry a negative charge. Clarification involves any or all of the following steps that cause particles to agglomerate:

- Introduction of cations such as aluminum or iron that are attracted to negatively charged particle surfaces.

- Introduction of a water-soluble, ionizing organic compound. A polymer, with numerous ionized sites per molecule, is preferred. These polymer cations, attracted to negatively charged particle surfaces, neutralize surface charges, and allow the particles to coagulate.

• Introduction of organic cations that hydrolyze to form insoluble precipitates and to entrap particles. Examples are aluminum or iron salts that form precipitates of either aluminum or iron hydroxide.

In most clarification operations, flocculation follows coagulation. Flocculation starts when neutralized or entrapped particles begin colliding and growing in size. This process occurs naturally but is enhanced by addition of coagulant aids.

Most inorganic coagulants are acid salts that lower the pH of treated water. Depending on initial raw water alkalinity and pH, often an alkali, such as lime or caustic, must be added to counteract pH depression of the primary or inorganic coagulant.

Since pH can affect both particle surface charge and floc precipitation during coagulation, it is an extremely important parameter. Iron and aluminum hydroxide flocs are best precipitated at pH levels that minimize hydroxide solubility. The best pH for coagulating suspended solids, however, is not always coincident with the pH for minimum hydroxide floc solubility.

The term polyelectrolyte refers to water-soluble organic polymers used for clarification, whether they function as primary coagulants or as flocculants.

The cationic polyelectrolytes commonly used as primary coagulants are described as polyamines. When used as primary coagulants they adsorb on turbidity surfaces and neutralize repelling negative charges. The polymer may bridge, to some extent, from one particle to another. This accomplishes clarification and flocculation without a requirement for precipitation of suspended solids by inorganic coagulant feed. The pH of the treated water is unaffected.

In certain instances, an excess of primary coagulant, whether inorganic, polymeric, or a combination of both, is fed to promote large floc size and settling rate. In some waters, however, even large doses of primary coagulant will not produce a satisfactory floc. A polymeric coagulant aid added after the primary coagulant can eliminate waste of a primary coagulant by developing a dense floc at low treatment levels.

Coagulant aids have proved successful in cold lime softening and clarification to achieve both improved settling rates of precipitation products and finished water clarity. The floc-enlarging process is called flocculation to distinguish it from coagulation.

Although the most efficient sequence for adding coagulation chemicals varies from one system to another, it is usually:

Chlorine,
Bentonite,
Primary inorganic and/or polymer coagulant
pH-adjusting chemicals, and
Coagulant aid
Polymers

Chlorine assists coagulation by destroying organic contaminants that have dispersing properties. Some water with high organic content has increased primary coagulant demand. Chlorination, prior to primary coagulant feed, will reduce the coagulant dosage. When using an inorganic coagulant, adjusting pH prior to the coagulant establishes the correct environment for the primary coagulant.

All treatment chemicals, except coagulant aids, should be added during turbulent mixing of the influent water. Rapid mixing while the aluminum and iron coagulants are added insures uniform cation adsorption onto the suspended matter.

Colored surface waters occur where drained ground areas contain excessive quantities of decaying organic materials. Swamps and wetlands introduce color into streams, particularly after heavy rainfalls. Color indicates various problems such as objectionable taste, increased bacteria and algae growth, fouling of ion exchange resins, interference with coagulation, and stabilization of soluble iron and manganese.

Most organic color in surface waters is caused by negatively charged colloids. Chemically, organic color-producing compounds are classified as humic and fulvic acids. Color can be removed by chlorination and coagulation with aluminum or iron salts. Chlorine will oxidize color compounds while inorganic coagulants will remove many types of organic color by neutralization of surface charges. Additional color removal is obtained by chemical interaction with aluminum or iron hydrolysis products. Cationic organic polyelectrolytes also can coagulate some types of color particles.

Optimum coagulation for color reduction normally is carried out at pH 4.5 to 5.5. Optimum pH for turbidity removal is usually much higher than for color reduction. Sulfate ions can interfere with color-reduction coagulation; whereas calcium and magnesium ions can improve the process and broaden the pH range in which color can be effectively removed.

Clarifier Configurations

The coagulation-sedimentation process requires three distinct operations:

- rapid mixing for coagulation,
- moderate mixing for flocculation, and
- floc and water separation.

Originally, conventional clarification units consisted of large rectangular concrete basins divided into two or three sections. Each stage of the clarification process took place in one section of the basin. Water flow was horizontal through the system. A horizontal settling basin is shown schematically in Figure 4-5.

Another type of clarifier is the upflow design. Compact, relatively economical, upflow clarifiers accommodate coagulation, flocculation, and sedimentation in a single, steel or concrete tank. These clarifiers are termed upflow because the water flows up while suspended solids settle. A key feature in maintaining a high clarity effluent is increased solids contact through internal sludge recirculation.

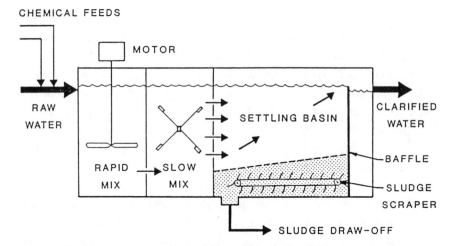

Figure 4-5. Horizontal Settling Basin

Because retention time, in an upflow unit, is less than in a horizontal system, upflow basins can be much smaller than horizontal basins of equal capacity.

Most upflow designs are either sludge blanket or solids-contact clarifiers. After coagulation and/or flocculation in sludge blanket units, incoming water passes through a suspended layer of previously formed floc. Figure 4-6 shows an upflow sludge blanket clarifier.

Another clarifier is the solids-contact design. Solids-contact refers to units in which large volumes of sludge are circulated internally. The term solids-contact also describes the sludge blanket unit and simply means that prior to and during sedimentation, chemically treated water contacts previously coagulated solids.

Solids-contact units often combine both clarification and precipitation softening. Bringing incoming raw water into contact with recirculated sludge improves efficiency of softening reactions and increases both size and density of floc particles.

Raw water analysis alone is not very useful for predicting coagulation conditions. Coagulation chemicals and appropriate dosages must be selected, either by operating experience with a given raw water or by simulation of the clarification step in a laboratory.

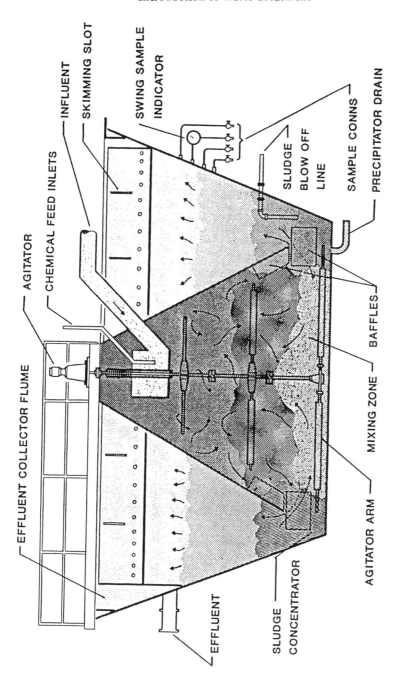

EFFLUENT COLLECTOR FLUME

AGITATOR

CHEMICAL FEED INLETS

INFLUENT

SKIMMING SLOT

SWING SAMPLE
INDICATOR

SLUDGE
BLOW OFF
LINE

SAMPLE CONNS

PRECIPITATOR DRAIN

BAFFLES

MIXING ZONE

AGITATOR ARM

SLUDGE
CONCENTRATOR

EFFLUENT

Figure 4-6. Upflow Sludge Blanket Clarifier

Filtration

Coagulation and sedimentation of a surface supply produces water of suitable quality for most industrial uses. When water is used for drinking, boiler makeup, or process cooling, additional particulate matter usually must be removed. Water can be filtered by passing it through a porous bed of sand or similar material.

Filtration is a mechanical process for removing suspended matter from water by passing it through a porous medium. Filtering does not remove dissolved solids, but it can be used together with a softening process that does reduce concentrations of dissolved solids. Sand or anthracite bed filtration, for example, usually follows the settling of precipitated hardness salts resulting from lime/soda ash softening.

In most water clarification or softening processes where coagulation and/or precipitation occur, at least a portion of the chemically processed water is filtered. Rapid filtration can markedly reduce turbidity of water that has been pretreated with either inorganic or synthetic-organic coagulants. If filtration follows coagulation, excessive suspended solids are not discharged into finished water distribution systems when upsets occur in sedimentation basins.

The flow in conventional rapid filters is downward. Both gravity and pressure filters usually operate this way. Sand or anthracite is usually the filter medium in a bed consisting of either one or two grades of sand or anthracite, with a total depth of 15 to 30 inches.

A gravel bed supports the filter medium, prevents fine sand or anthracite from passing into the underdrain system, and distributes backwash water. The supporting bed consists of 1/8 to 1½ inch gravel in graded layers to a depth of one to two feet.

Sand, anthracite coal, calcite, garnet, magnetite, and other materials may be used for filtration. Silica sand and anthracite are most commonly used. When a siliceous medium is not suitable, such as in a filter following a hot process softener where the treated water is intended for boiler feed, anthracite is usually used.

The size and shape of filter particles govern efficiency. Sharp, angular particles form large voids and remove less fine material than rounded particles of equivalent diameter. The filter medium must be coarse enough to allow sludge to penetrate the bed for two to four inches. Although most suspended solids are trapped either at the surface or in the first two inches of bed depth, some penetration is essential to prevent rapid loss of head.

Multi-layer, in-depth, reverse-graded, and mixed media are terms applied to a type of filter bed that is size and density graded so that the coarsest, least-dense medium is at the top and the finest, densest medium is at the bottom. Downflow filtration through this filter allows deeper and more uniform penetration by particulate matter, and permits higher filtration rates and longer runs. Because of density differences of the different media, layer configuration is maintained even after high-rate backwashing.

Replacing the top portion of a rapid sand filter with anthracite is termed capping. From two to six inches of 0.4 to 0.6 mm sand are removed from the surface of a bed and replaced with approximately four to eight inches of 0.9 mm anthracite. If the reason for capping is to increase capacity, a larger amount of sand is replaced. Pilot tests should be run to assure that reducing the depth of the finer sand layer does not reduce quality of the effluent.

The essential parts of a gravity filter, apart from the filter medium, are:

- Filter Shell: may be concrete, steel, or wood and may be square, rectangular, or circular. Rectangular reinforced concrete units are most widely used.

- Gravel Bed: supports the filter medium and prevents loss of fine sand or anthracite to the underdrain system. The supporting bed, usually one to two feet deep, distributes backwash water.

- Underdrain System: supports the gravel layer and also assures uniform collection of filtered water and uniform distribution of backwash water. The system may consist of a header and laterals, with perforations or strainers suitably spaced. False bottom underdrain systems are also used.

- Wash Water Troughs: sized large enough to collect backwash water without flooding. The troughs are spaced so that horizontal travel of backwash water does not exceed 3.5 feet. In conventional sand bed units, wash troughs are placed approximately two feet above the filter surface. Sufficient freeboard must be provided to prevent loss of a portion of the filter medium during maximum backwash rates.

- Control Devices: assure maximum efficiency of filter operation. Rate-of-flow controllers, that operate from venturi tubes in the effluent line, automatically maintain uniform delivery of filtered water. Backwash rate-of-flow controllers are also used. Rate-of-flow and loss-of-head gauges are essential for efficient operation. Operating controls and gauges for either manual or automatic operation are normally grouped on an operating table for convenience.

Placing pressure filters in-line eliminates the need for pumping filtered water. Pressure filters are used with hot process softening to permit high-temperature operation and to prevent thermal loss. Pressure filters, like gravity filters, have a filter medium, supporting bed, underdrain system, and control devices. The filter shell differs from a gravity filter shell as it has no wash water troughs.

Upflow filtration passes water up through a single medium, graded bed. It has some of the advantages of coarse-to-fine filtration, and several drawbacks not encountered in downflow, mixed media filtration.

The major disadvantage of upflow filtration is the difficulty in washing. Downflow washing of an upflow filter produces neither bed expansion nor scouring action. Upflow washing, although effective, can leave a portion of the coarse dirt particles trapped in the lower layers of the filter media. Another major drawback is that excessive bed expansion during operation limits the maximum flow rate for satisfactory filtration.

Several manufacturers have developed gravity filters that backwash automatically at a preset head loss. Head loss actuates a backwash syphon that draws wash water from storage, up through the bed, and out through the syphon pipe. A low level in the backwash storage section breaks the syphon; the filter then rinses and returns to service.

Filters must be washed periodically to remove accumulated sludge. Inadequate cleansing allows the formation of permanent clumps, in increasingly larger areas, and filtering capacity gradually decreases. If fouling is severe, the medium must be removed from the unit and either chemically cleaned or replaced.

Rapid downflow filters are washed by forcing clear water back up through the media. In conventional gravity units, backwash water lifts solids from the bed into wash channels and carries them to waste. Where only water is used for backwash, the backwash is normally preceded by surface washing, wherein strong jets of high-pressure water from fixed or revolving nozzles break up surface crust. Water backwashing usually takes 5 to 10 minutes. After backwashing a small amount of rinse water is filtered to waste and the filter is returned to service.

High-rate backwash can cause mud balls to form inside the filter media. An excessive backwash rate and resultant bed expansion can produce random currents causing some zones in the expanded bed to move downward while other zones move upward. These currents can carry sludge-encrusted surface media deep into the bed where swirling action of currents forms mud balls. Efficient surface washing helps prevent this.

Air scouring with low-rate backwashing can break up surface crust without producing random currents, but the underdrain system must be designed to distribute air uniformly. Solids removed from the medium collect in the layer of water between the medium surface and the wash channels. After air is stopped, dirty water is normally flushed out, either by increasing the backwash water flow or by draining it from the surface. Wash water consumption is about the same whether water only or air/water backwashing is used.

Removing coagulant-treated suspended solids from water by rapid filtration is called in-line clarification, in-line filtration, or contact filtration. It may follow coagulation alone or coagulation and partial sedimentation. The process removes colloidal-size solids with no need for large sedimentation basins.

The kinds of filters that can be used for in-line clarification depend on the coagulant chemicals used. Generally, one of the following coagulation schemes is used:

- An inorganic aluminum or iron salt coagulant, alone or combined with either an alkaline chemical to correct pH of the water or a high-molecular-weight polymeric coagulant aid.

- A strongly cationic organic polyelectrolyte.

Microfiltration

Microfiltration is used to remove extremely small particulate matter, oil particles, and even bacteria. This method is practical only for relatively small quantities of water containing low concentrations of removable contamination. Diatomite precoat filtration is the most common and most effective of these processes.

Microfiltration is used industrially to polish water already clarified by conventional means, resulting in extremely pure water. Often precoat filters are used to remove oil from contaminated condensate. The U.S. Army uses portable diatomite filters to provide potable-grade water in isolated areas.

In the precoat filtration process, the precoat of diatomaceous earth acts as the filter medium and forms a cake on a permeable base (Figure 4-7). The base or septum must prevent passage of diatomaceous particles, without restricting the flow of filtered water, and must withstand the relatively high pressure differentials produced. Substances used as base materials are filter cloths, porous stone tubes, porous papers, wire screens, and wire wound tubes.

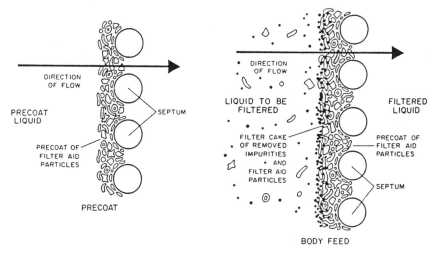

Figure 4-7. Principles of Diatomite Filtration
Courtesy of Dow Chemical Corporation, April 1991

The supporting medium is first coated with a slurry of diatomaceous earth. Water is filtered by the diatomite coating. Additional diatomite slurry is usually added during the filter run. When the accumulation of matter removed by filtration generates a high backpressure, the filter coating is sloughed off by backwash. The filter bed is then recoated and returned to service. Chemical coagulants are usually not required, but have been used where an ultrapure effluent is needed.

Silica mineral deposits formed by diatoms (common name for minute planktonic unicellular or colonial algae) are the sources of diatomaceous earth. The deposits are composed of billions of tiny diatom skeletons. Commercial filter aids are produced from the crude material by a gentle milling process which yields a fine powder of rigid particles so minute that hundreds could be placed upon a pinpoint. Precoat requirements are generally from one to two ounces per square foot of filter area. Continuous or body dosages are directly influenced by influent turbidity and by filtration rates, and vary widely. Values ranging from 5 to 50 ounces per thousand gallons of filtered effluent have been reported.

While diatomite is the filter aid primarily employed in precoat filtration, various grades of purified wood cellulose are used to filter high pH solutions where silica pickup cannot be tolerated. Other media include perlite, asbestos, and carbon. Perlite is unacceptable when silica control is necessary.

Note that microfiltration is also possible with various types of porous membranes. This will be discussed further under "Membrane Technology."

Activated Carbon Adsorption

Carbon adsorption is a physical process in which a contaminant is transferred from an aqueous or vapor phase to the surface of the solid carbon, where it accumulates for subsequent extraction or destruction.

Some dissolved metals and inorganic species, including arsenic, mercury, silver, chromium and chlorine, also have shown good potential for being adsorbed by carbon. Although carbon absorption rates for many individual species in aqueous and vapor phases can be estimated from existing experimental data or empirical equations, column studies are required to evaluate an overall adsorption rate if several species are present and competing for the activated carbon's adsorption sites. Once an overall adsorption rate is measured, equipment can be sized and designed.

In some cases, carbon filters have been applied after the reverse osmosis (utilizing a cellulose membrane). Inasmuch as reverse osmosis will usually remove most of the bacteria in the system, follow-up with a carbon bed may defeat the work of the RO unit by introducing bacteria. Any residual disinfectant remaining in the RO effluent may be removed by a reducing agent such as sodium bisulfite, although the sulfite will slightly increase the TDS in the system. On the other hand, carbon support equipment such as ultraviolet devices and cartridge filters will not be required.

Generally, when ultra filters are used after a particulate filter, the ultra-filtration will remove the heavy molecular weight constituents. These are not effectively removed by carbon filtration. The lighter weight molecular constituents are not effectively removed by ultra-filtration, but will be taken out by the carbon filter.

Activated carbon filters are classically used to clean up water which contains chloramines, volatile organic compounds, and other compounds which are difficult to remove by conventional filtration. Carbon filters should be positioned in the process stream after the particulate filters, but before the ion exchange or reverse osmosis treatment. Carbon filtration serves an extremely important function in removing chlorine materials. Reverse osmosis and ultra-filtration membranes are also very sensitive to trace concentrations of organics or disinfectants. Activated carbon filtration is also useful in removing trace concentrations of various organic and inorganic compounds, possibly existing in high purity electronic processes. Many potable water filters contain carbon to remove trace chemicals. A danger exists, however, in the use of a carbon filter: the bed is a favorable breeding ground for bacteria, which will multiply rapidly in this favorable environment. Subsequent bacterial destruction may be achieved by ultraviolet or the use of ozone. Because iron acts as a nutrient for bacteria, iron should not be used as pipe and valve material if bacteria growth is a concern.

As a purification step, carbon adsorption is rarely used alone. It is especially effective for protecting ion exchange resins from chlorine attack and organic fouling. Since carbon removes free chlorine from water, bacteria may grow downstream to undesirable levels. As a result, the placement of carbon beds in relation to other components is an important consideration in system design.

TREATMENT METHODS

The purpose of treatment methods used to remove or modify dissolved solids and gases in water is the prevention of scale and corrosion. The general techniques are chemomechanical, ionic manipulation, or some combination of both.

The methods discussed in increasing order of sophistication (and the order in which they should be evaluated to solve a water problem) are:

Media Filtration
Deaeration
Carbon Filtration
Evaporation
Lime (hot or cold)
Sodium Zeolite Softening
Electrodialysis
Reverse Osmosis
Dealkalizing
Desilicizing
Demineralization

Selection of the type of treatment is dependent on the end use of the water. The cost of the treatment methods varies greatly. To achieve the desired result at a practical cost, several processes may be operated in parallel, or, more commonly, in series.

Table 4-3 summarizes results achieved by various methods of boiler feedwater treatment.

Deaeration

Water containing dissolved gases causes many corrosion problems. For instance, oxygen in water produces pitting that is particularly severe because of its localized nature. Carbon dioxide corrosion is frequently encountered in condensate systems and, less commonly,

Table 4-3. Results of Various Water Treatment Processes
for Boiler Applications
(all concentrations expressed in parts per million)

Boiler Pressure Range, psi	External Equipment Employed	Total Hardness	Resultant Feedwater		Total Solids
			Alkalinity	Silica	
0-300	Sodium Zeolite	0-2	No change	No change	No change
	Split-Stream Softening	0-2	15-20	No change	Reduced with Alkalinity
	Sodium Zeolite/- Chloride Anion Dealkalization	0-2	10% Raw Water	No change	No change
	Hot Process Softening	10-20	40-60	1-2	Reduced with Alkalinity
300-900	Hot Process/Hot Zeolite	0-0.1	20-40	1-2	Reduced with Alkalinity
	Demineralization	0-0.1	0-5	0-1.0	0-5
900 & above	Demineralization	0-0.1	0-5	0-0.1	0-5
	Evaporation	0-0.5	0-10	0-0.1	0-10

Courtesy of Betz Laboratories

in water distribution systems. Water containing ammonia readily attacks copper and copper-bearing alloys. Hence, deaeration is widely used to remove dissolved gases from water to control corrosion. In particular, it is used to remove oxygen from boiler feedwater systems. While conventional power plant deaerators remove many dissolved gases from water, frequently they do not remove the large quantities of dissolved gas found in many industrial plant waste streams.

Pressure deaeration, with steam as the purge gas, is used to prepare boiler feedwater. Steam is chosen as purge gas because it is readily available, it heats the water and reduces solubility, and it does not contaminate the water. Only a small quantity of steam need be vented, because most of the steam used is condensed into the deaerated water.

Pressure deaerators, used to prepare boiler feedwater, produce deaerated water that is very low in both dissolved oxygen and free carbon dioxide. Vendors usually guarantee less than 0.005 cubic entimeters (cc) per liter of oxygen and zero free carbon dioxide whenever the bicarbonate alkalinity exceeds five ppm.

Boiler feedwater is deaerated by spraying or cascading the water through a steam atmosphere. This heats the water to within a few degrees of the temperature of the saturated steam. Since solubility of oxygen in water is very low under these conditions, 90 to 95 percent of the oxygen in the incoming water is released and purged from the system by venting. The remaining oxygen is not soluble under equilibrium conditions, but it is not readily released. The water leaving the heating section of the deaerator, therefore, must be vigorously scrubbed with steam to remove the last traces of oxygen. Some gases are not as readily removed from water as oxygen.

Since only the ammonium hydroxide form exerts a gas solution pressure which permits ammonia to be removed by deaeration, it is most efficiently removed at alkaline pH levels. With the best ammonia removal achieved at a high pH, and the best carbon dioxide removal at a low pH, complete degasification of a stream containing a combination of the two is difficult to achieve by deaeration.

The two major types of pressure deaerators, or deaerating heaters, are the tray-type and the spray-type.

A tray-type deaerating heater, shown in Figure 4-8, consists of a shell, perforated inlet water spray distribution pipes, direct contact vent condenser, tray stacks, and protective interchamber walls. While the shell is of low carbon steel construction, more corrosion-resistant stainless steels are used for distribution pipes, vent condenser, trays, and interchamber walls.

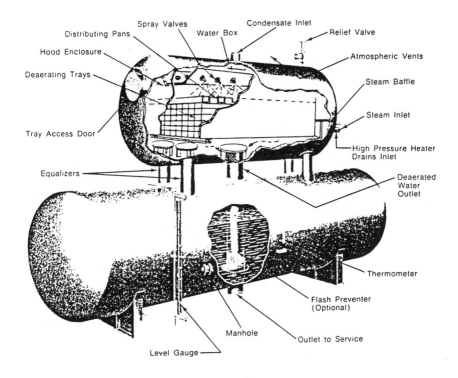

Figure 4-8. Tray-Type Deaerating Heater

Source: BELCO

In central utility power generating stations, main turbine condensers have air ejectors that remove dissolved gases. Sometimes, because deaeration is obtained from these condensers, the pressure deaerator is omitted from the feedwater cycle. There is a danger of air leaking into the system, both during start-up and while the condensers are operating at low loads; however, pressure deaerators are more often included in central station designs.

Vacuum deaeration is used for deaerating water at temperatures below the atmospheric boiling point to reduce the corrosion rate in water distribution systems. The vacuum deaerator, like the pressure deaerator, is designed to bring water to its saturation temperature (by applying a vacuum to the system rather than heating the water), to divide the water into fine droplets to facilitate gas removal, and to vent the gases from the unit.

Pressure deaerators, used to prepare boiler feedwater, produce deaerated water that is very low in both dissolved oxygen and free carbon dioxide. Vendors usually guarantee less than 0.005 cubic centimeters (cc) of oxygen per liter and zero free carbon dioxide whenever the bicarbonate alkalinity exceeds five ppm.

Vacuum deaerators, used to protect water distribution lines, are not designed to provide as complete deaeration as pressure deaerators. They will usually reduce dissolved oxygen to a range of 0.25 to 0.50 cc per liter.

Deaerators function efficiently with minimum maintenance and are usually economical to operate.

Inlet water to deaerators should be free of suspended solids that could clog spray valves and ports of the inlet distributor and the deaerator trays. In addition, spray valves, ports, and deaerator trays may plug with scale that forms when the water being deaerated has high hardness and alkalinity levels.

While pressure deaerators reduce oxygen to low levels, even trace amounts of oxygen may cause corrosion damage to a system. Consequently, good operating practice requires removal of the last traces of oxygen with a chemical scavenger. Sodium sulfite or hydrazine are commonly used, although many other compounds are also useful.

Free carbon dioxide can be removed by deaeration, but the process releases only small amounts of combined carbon dioxide. The majority of the combined carbon dioxide is released with steam in the boiler and subsequently dissolves in the condensate, causing corrosion problems.

Alternate Means of Deaeration

Several other methods of deaerating water have been developed and are now available. These include (1) removal by using ion exchange techniques; (2) the Echolochem system; and (3) removal by using a catalyst and hydrazine.

A paper by Breslin and Adams, presented at the International Water Conference of the Engineers' Society of Western Pennsylvania in October of 1961 first proposed the use of ion exchange resins for the deaeration of water. These techniques have been further developed, to the extent that a 1990 proposal by a chemical company reads, in part, as follows:

> Oxygen scavenging by ion exchange is carried out by using a Strong Base Anion (preferably Type I) in the sulfite form. Dissolved oxygen in condensate which is normally in the low ppm level is easily removed to nontraceable levels. TULSION A-23, Strong Base Anion Exchange Resin, Type I, is regenerated with an alkaline mixture of Sodium Sulfite and Sodium Hydrosulfite. Dissolved oxygen in the 1.0–2.0 ppm ranges is easily reduced to nontraceable levels, under standard operating flow rates. Lower flow rates are essential to remove dissolved oxygen in the 3–10 ppm ranges to nontraceable levels.
>
> You could also consider TULSION A-27 MP, Macroporous Strong Base Anion, Type I. Thermax's macroporous resins offer the unique superior operating characteristics of their gel equivalents with higher physical stability and resistance to organic fouling.

Thermax states that one liter of their resin will remove six grams of air, prior to regeneration. It is evident therefore that although this technique is not well known, it is a practical way of removing air from water, in cases where a deaerator is not feasible.

Echolochem, engaged in mobile water treatment services, has developed an exclusive, proprietary deoxygenation process utilizing their "MobileFlow Water Treatment System." This method employs the use of hydrazine, with passage through a bed of activated carbon to catalyze the reaction, followed by ion exchange to remove any carbon impurities or hydrazine overfeed. Reactions are completed in seconds. Production of 10 ppb of dissolved oxygen levels is attained immediately, and can often be maintained for a period of several hours after shutdown of the hydrazine feed, apparently due to a "flywheel effect."

The Echolochem Deox Process is claimed to dramatically reduce dissolved oxygen in makeup water. Deoxygenation is combined with ion exchange polishing to produce up to 600 gpm of demineralized water which contains extremely low levels of dissolved oxygen, typically measured at less than ten parts per billion (ppb).

Westinghouse has developed a Catalytic Oxygen Removal System (CORS), now marketed by Ionics, which has promise for specific uses. The process operates as follows:

> The catalytic oxygen removal system (CORS) uses a palladium doped anion resin to catalyze the reduction of dissolved oxygen in water or other aqueous solutions with either hydrogen (H_2) or hydrazine (N_2H_4). The reaction takes place at ambient temperatures and a pressure of approximately 100 psig. The dissolved oxygen level can be reduced in a single pass from the 6.0 to 8.0 ppm range down to less than 10 ppb, less than 5 ppb, or less than 1 ppb—depending on design. The reaction by-products are water for the hydrogen system and nitrogen for hydrazine. The anion resin acts only as a carrier for the palladium catalyst and is not regenerated. The resin bed may need occasional backwashing depending on the suspended solids level of the solution being deoxygenated. CORS units have been designed in the range of 100 to several thousand gallons per minute. The system will operate using either hydrogen, or hydrazine, as preferred.

Figure 4-9 shows the process.

Iron Coated Sand—A New Technique (Reported November 1990)

University of Washington engineers have developed a treatment technique using sand coated with iron minerals to remove dissolved and particulate toxic metals from industrial wastewater

Lab tests also indicate the process may be useful for domestic water supplies. The coated sand traps organic materials that form

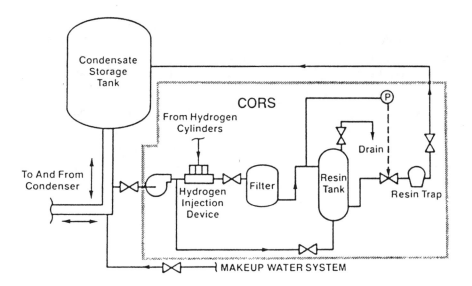

Figure 4-9. Schematic of the CORS Process

trihalomethanes when water is chlorinated. Trihalomethanes are suspected carcinogens.

The treatment was invented by Mark Benjamin, professor of civil engineering in UW's environmental engineering and science program, and two graduate students, Marc Edwards and Paul Anderson. Patents have been applied for.

The treatment relies on the chemical process of adsorption to cause dissolved metals such as lead, cadmium, copper, nickel and chromium to attach to the iron on the sand's surface. Iron oxide adsorbents have been used previously to process metal-bearing wastewater, but large amounts are required and they present a disposal problem, according to Benjamin.

The engineers found a way to coat sand with a tiny layer of an iron oxide called ferrihydrite. The coated sand is then packed in a column. Wastewater flows through the column and the coated sand adsorbs metal wastes. There is a buildup of metals over time, but the wastes can be collected by reversing the filtration process. A regenerating solution is passed through the coated sand and contaminants are collected in a highly concentrated form, in a solution occupying about one-half of a percent of the volume of the water treated. The coated sand can be reused.

"We have regenerated the sand 40 times, and the material was working without a noticeable loss in treatment efficiency," said Benjamin. "There is some loss of iron over time, so we don't know yet the maximum number of times we can use the coated sand."

The American Water Works Research Foundation is funding a two-year series of tests at the Everett city water treatment plant to evaluate the system's ability to remove organic material from drinking water. Meanwhile, a grant from the Environmental Protection Agency will permit further research in treating industrial wastewater.

Evaporation

Evaporation is used to remove solids. There are many types of evaporators, and several factors govern their selection. These include ease of operation, control of scale formation, accessibility for cleaning, and required purity of water. Evaporator systems commonly are employed aboard ships.

In an evaporation system, water to be treated is heated in a chamber and converted to vapor, leaving behind impurities in the unvaporized water. The vapor is then condensed back to a liquid. The source of heat used to vaporize the water typically is either high or low pressure steam. A simple schematic of an evaporator system is shown in Figure 4-10.

A more elaborate evaporator, using the principle of Vapor Compression, is shown by Figure 4-11. This system requires heat to start it up, but most of the driving force for the evaporation is provided by the process itself. This system is popular on ships, and where the supply of heating steam is severely limited. The evaporation can also

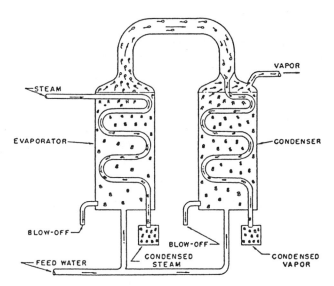

Figure 4-10. Elementary Two Stage Evaporator

be sustained by exhaust heat from an engine, so the device can be used wherever engine heat can be recovered.

The evaporation process can occur at any temperature corresponding to the boiling point of water at the pressure in the chamber. Accordingly, some evaporators operate below 212°F, the standard boiling point of water at atmospheric pressure.

Because evaporators require large amounts of energy to operate, they have lost favor, although they are still being used in appropriate circumstances.

Several types of evaporators have been developed. They are classified as submerged tube, film, flash, and compression evaporation. They can also be arranged as single, double, or multiple effect. High efficiency evaporators used for water making in the Saudi Arabia area have as many as thirty effects.

An evaporator consisting of a single heating chamber is a single effect unit. For economy, it may be desirable to arrange a number of units in series. They are then referred to as "multiple effect" units.

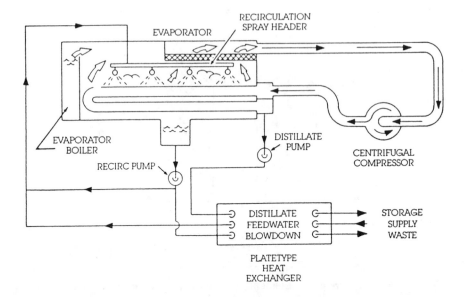

The feedwater is fed through a plate type heat exchanger, and preheated by the outgoing distillate and blowdown. It enters the brine recirculation header and mixes with the recirculated brine. The recirculated brine is sprayed over the evaporator tube bundle and collected in the evaporator sump. The brine is then pumped back into the recirculated brine header, with a portion of the recirculated brine being discharged to the heat exchanger, cooled, and sent to waste. The discharged brine, or blowdown, will control the recirculated brine concentration. The starting and make-up steam is produced in the evaporator boiler. Steam is generated from the recirculating brine as it is sprayed over the heat transfer surface of the tube bundle. The steam is drawn through the demisters to the centrifugal compressor which increases the pressure and temperature of the steam and discharges it into the heat transfer tube bundle, where it condenses into distillate. The steam generated by the process is pure water vapor; therefore, when it is condensed it is pure distillate. The distillate is collected at the bottom of the steam chest and is pumped through the heat exchanger, cooled, and discharges to storage. Under clean tube conditions, the unit produces more than its rated capacity which slowly decreases as scaling takes place. Scale formation is inhibited by the chemical treatment injected. Cold seawater flushing during normal equipment shutdown periods increases the operating time between acid cleanings. After an extended period, the unit is acid cleaned to remove all scale and begin the cycle again with over-capacity.

Figure 4-11. Spray Film Vapor Compression

In a series arrangement, the vapor from one effect is used to heat and vaporize water in the next unit of the series. The vapor coils of each effect then serve as the condenser for the previous effect. By using the vapor generated in each successive effect, the efficiency of the assembly is increased. A reduced quantity of primary steam is required to evaporate a given amount of feedwater. Figure 4-11 describes a more complex arrangement.

The amount of distilled water produced by an evaporator per pound of steam used is the "evaporator economy." Single-effect evaporators will deliver about 0.9 pounds of distillate per pound of steam. Typical evaporator performance for multiple-effect units is as follows:

	Ratio of Pounds of Steam Supplied/Pounds of Distillate Produced
Single-effect evaporator	1/0.9
Double-effect evaporator	1/1.65
Triple-effect evaporator	1/2.3
Quadruple-effect evaporator	1/2.8
Twenty-effect evaporator	1/6 to 1/10

Evaporators may be obtained with performance guaranteed to produce a distillate that contains no more than 0.5 ppm total solids when fresh water is used for evaporator feed. If sea water is used as feed, the guarantee may warrant that distillate will contain no more than 0.25 grains (4.0 ppm) of total salts per gallon.

Carryover from an evaporator, similar to carryover from any low-pressure boiler, consists of vapor in which the shell water is either chemically or mechanically entrained. The concentration of shell water is, perhaps, the most important influence on vapor purity. A limiting concentration may be specified by the evaporator manufacturer.

Antifoams may be used to improve steam purity the same way they are used in boilers. Depending on the characteristics of the feedwater, chemical treatment may be needed to prevent scale formation.

Raw water containing high concentrations of solids can be fed directly to evaporators, but performance and maintenance can be improved by pretreatment of the evaporator feed.

Pretreatment by sodium zeolite softeners eliminates practically all scale-forming solids from the evaporator. The total soluble solids in zeolite softened water is somewhat greater after treatment than before. It is desirable, however, to soften makeup water to the evaporator both to reduce outage periods and to increase efficiency.

Deaeration of feedwater to evaporators is desirable for reducing corrosion both in the evaporator and in the condensate lines. Oxygen can be eliminated completely, and free carbon dioxide can be reduced to low levels. Carbon dioxide carried over with vapor is dissolved when the vapor is condensed, and because the condensed liquid contains few buffering solids, the water may be acidic.

It is characteristic of evaporators that capacity falls off as scale builds up on the tubes.

In practice, units operating on sea water require scale cleaning every 300 to 700 hours. With fresh water, this period is extended to as much as 2,000 or 3,000 hours.

Some evaporators are designed for "cold shock" treatment. When the evaporator is shut down but still hot, cold water is sprayed on the hot tubes. The rapid temperature change causes tubes to contract suddenly, and adherent hard scale cracks off.

Lime—Hot or Cold (Figure 4-12)

Use of lime to remove hardness probably is the oldest chemical process for water treatment. The first plants built for this type of treatment were merely large tanks where hard water was mixed with lime water. The mixture was allowed to stand and precipitates formed and settled. Clear water was decanted by a swing-pipe, and

Flow Diagram for Cold Lime — Zeolite Process

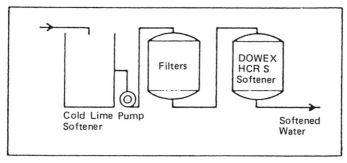

Flow Diagram for Hot Lime — Zeolite Process

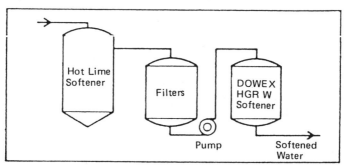

Figure 4-12. Simplified Schematic of Hot and Cold Lime Processes
Courtesy of Dow Chemical Company, April 1991

filtered. Similar equipment, with slight modifications, is used for batch softening today.

In the cold lime-soda process, certain impurities are removed from water by chemical precipitation. Chemicals are added that form insoluble compounds, which are allowed to settle. The relatively clear, settled water is filtered to remove the last traces of solid matter.

Aside from its operational problems, the hot lime system has the great advantage of removing both high levels of silica and hardness. Hot lime zeolite may be considered when silica and hardness levels are very high.

The cold lime-soda process, serves two purposes—removal of hardness and reduction of alkalinity. In this process three reactions take place:

- Calcium hardness is converted to insoluble calcium carbonate (limestone)

- Magnesium hardness is converted to magnesium hydroxide

- Alkalinity reduction is obtained.

The cold batch process, to a large extent, has been supplanted by the hot continuous process.

One standard means of treating boiler makeup feedwater is by the hot precipitation process. This method, while reasonably effective, has certain disadvantages. Significant space is required for equipment, and elaborate chemical feed controls must be maintained to meter the lime, soda ash, and, occasionally, phosphates. In addition, both soda ash and phosphates are relatively expensive.

A combination of ion exchange and hot lime softening solves most problems inherent to the cold process, while offering economic benefit.

In the treatment of water by hot-lime-soda followed by phosphate, the most common hot precipitation method, lime and soda ash are added first to the raw water supply. Lime reacts with calcium bicarbonate and soluble magnesium salts in the water, precipitating insoluble calcium carbonate and magnesium hydroxide.

The function of soda ash, which is generally added in excess, is to precipitate, as carbonates, soluble calcium salts from the reaction, as well as carbonates occurring naturally in the water.

Generally, hardness is reduced to about 1.5 grains per gallon by addition of approximately 1.5 grains per gallon excess soda ash. For many purposes, however, a hardness of 1.5 grains per gallon is objectionable. Consequently, mono- or di-sodium phosphate is added in the second stage of treatment. The reaction with phosphate lowers hardness to a few parts per million, but additional carbonate alkalinity is introduced. Thus, the steam produced may contain considerably more carbon dioxide than desirable.

As noted, soda ash and phosphates are relatively expensive. It is apparent that to approach zero hardness, an excess of phosphate must be present. Unless the pH of the water is maintained at about 8.5 there is danger that calcium phosphate may deposit in second stage heaters or in economizers. Thus addition of acid is necessary to maintain the pH at 8.5.

In the hot lime-ion exchange process, lime is added to the raw water. Expensive soda ash is not used. As in the hot lime-soda-phosphate method, insoluble calcium carbonate and magnesium hydroxide are precipitated. Hardness is reduced in proportion to the alkalinity of the raw water. The effluent, partially softened and of low alkalinity, is then filtered to prevent fouling of the resin bed with suspended carbonates, which may cause decreased capacity and premature breakthrough of hardness.

The lime-treated, filtered water is passed through the sodium form of a strongly acidic cation exchange resin, which scavenges residual hardness out of the water, yielding makeup water with essentially zero hardness. The amount of phosphate required to complete the job is very small. Since a high capacity resin is employed, it is able to handle large volumes of water having variable composition. Also, because of this capacity and because influent water from the lime softener is partially softened, the resin requires regeneration at infrequent intervals..

The primary advantages attained with the hot lime and zeolite softener method over the hot lime-soda-phosphate process are:

- Lower cost of lime, salt and the small quantity of phosphate required, compared with higher costs associated with the hot lime-soda-phosphate method

- Zero hardness attained

- Very low carbon dioxide content of steam produced with ion exchange treated water

- No need for elaborate controls
- Economy of space achieved
- Constant quality of effluent despite variations in influent.

The advantages of the hot over cold systems are higher rate of lime reaction, lower alkalinity and silica, and less corrosion due to near absence of oxygen.

Table 4-4 illustrates how each method changes the water. The "typical water" exhibited contains appreciable permanent hardness, and is especially suited to lime-ion exchange treatment.

The hot lime-zeolite process is actually very complex to operate and maintain. Contributing to the general unpopularity of this process today are high first cost, complexity of plant, and difficulty of maintaining continuous trouble-free operation.

Membrane Technology

Concurrent with the rapid rise of plastics technology since the last great war, the development of permeable membrane technology has led to new chemical separation methods which can actually segregate molecular elements by size. Such membranes are also known as "molecular sieves," since they act to separate molecular constituents of gases and liquids according to size. Separation of gases is a developing technology which is proving useful in separating and isolating elemental gases in various mixtures. Of more interest, however, in the field of fluid purification is the use of membranes for removing dissolved solids and gases from water sources, thereby providing utility managers one more option for developing clean water supplies from sources hitherto thought worthless.

Ultrafiltration, reverse osmosis, and electrodialysis have been available for some fifteen years. These methods have not achieved universal popularity in the U.S. because plentiful cheap water supplies in many areas have not justified the expense and complexity of membrane technology.

Table 4-4. Results of Treatment
(Cold Lime, Hot Lime, Processes)

Identification of Analysis Tabulated Below							
A. Typical raw water D. After hot lime-zeolite softening B. After cold lime softening C. After hot lime softening							
Constituent	PPM as	A	B	C	D	E	F
Calcium	CaCO₃	200	100	175	Nil		
Magnesium	CaCO₃	100	90	10	Nil		
Sodium	CaCO₃	100	100	100*	285*		
Total Electrolyte	CaCO₃	400	290	285*	285*		
Bicarbonate	CaCO₃	150	0	0	0		
Carbonate	CaCO₃	0	40	30	30		
Hydroxyl	CaCO₃	0	0	5	5		
Sulfate	CaCO₃	150	150	150*	150*		
Chloride	CaCO₃	100	100	100*	100*		
Nitrate	CaCO₃						
M Alk	CaCO₃	150	40	35	35		
P Alk	CaCO₃	0	20	20	20		
Carbon Dioxide	CaCO₃	5	0	0	0		
pH	——		7.5	9.5	10.2	10.2	
Silica	SiO₂	10	10	1-2	1-2		
Iron	Fe						
Turbidity							
TDS							
Color							

* These constituents may be reduced as much as 15% due to dilution by condensation of steam used in heating.

Courtesy of the Dow Chemical Company, March 1991

That picture is changing. Many of our traditional water sources are being overdrawn, while improvements in membranes have made them much less expensive to install and operate than was previously possible. As a result, they present a viable alternative to traditional treatment processes.

The membrane processes in current use may be unfamiliar to many people interested in water supplies for their processes. There are basically two types of separation processes: pressure driven, and voltage induced. The tabulation in Table 4-5 will illustrate the usual applications.

Table 4-5. Comparison of Membrane Process Design Features

Parameter	Pressure Driven Reverse Osmosis	Ultrafiltration	Voltage Driven Electrodialysis
species crossing the membrane	water	water	ions
species removed from water	all inorganics Fe and Mn may give trouble some heavy molecular weight organics nonreactive silica	colloidial suspended solids	ionized solids some organics
species not removed	gases (suspended solids are not considered since any quantity tends to block)	ionic and dissolved solids	gases, silica, non-ionized organics and particulates

Both types of processes utilize a common terminology:

Flux rate—The flow rate water, that passes through a unit area of membrane surface.

Permeate—The product water passing through the membrane. The term is often used to differentiate it from the product water that has received post treatment and has possibly blended with water from other sources.

Recovery—The percentage of the feedwater converted to product water.

$$Y = \text{product flow/feed}, \%$$

Concentration ratio—The concentration of the reject divided by the concentration of the feedwater, or approximately:

$$CR = 1/(1 - \text{recovery})$$

Rejection—The percent of dissolved substances that are left behind as the water permeates through the membrane.

$$R = \frac{cb - cp}{cb}$$

where cp = concentration of permeate and cb = concentration of brine.

The following tabulation generally classifies the uses of membrane technology:

Reverse Osmosis: Reverse Osmosis (RO), also called hyperfiltration, is capable of the highest filtration level possible, including separating dissolved salts and removing bacteria, pyrogens and organics from water. Cellulosic, polyamide and specialty polymer membranes are available in many varieties and configurations to meet a wide range of purification applications.

Nanofiltration: Not shown on Figure 4-13, as this is a very limited application. Nanofiltration (NF) equipment removes particles in the 300–1000 molecular weight range, rejecting selected salts, most organics and passing more water at lower pressure operation than RO systems. Nanofiltration economically softens water without the pollution of salt-regenerated systems and provides unique organics desalting capabilities.

Ultrafiltration: UF membranes provide filtration in the 0.0015 to 0.1 micron (1000 MW to colloidal) range. The feed fluid can be precisely fractionated, making UF ideal for fluid separations, industrial processes and as pretreatment or post-treatment for ultrapure water systems.

Microfiltration: Microfiltration (MF) membrane and equipment is used for demanding fluid separation and purification applications such as biotechnology and research. MF removes particles in the range of 0.05 micron to 2 microns in diameter. Solutions pass directly through the MF membrane in normal flow, or are separated into two effluent streams in a longer-lasting crossflow mode. Figure 4-13 is a chart of dimensions of various particles, and indicates the procedures favored for their removal.

Electrodialysis (ED)

Electrodialysis is an ion transfer process using ion selective membranes and small quantities of electrical energy to remove or concentrate dissolved salts in water. ED offers several unique advantages over classical desalting technology. Salt can be removed to specified levels rather than total salt removal, as in distillation, which often requires some mineral replacement for taste enhancement or corrosion control. Also, the electrodialysis process can adapt to changing mineral content of either feedwater supply or desired product water. Energy requirements for the electrodialysis process are in direct relation to the quantity of salts removed. Therefore, as salt removal requirements increase, so do energy costs.

Commercial units consist of at least one pair of electrodes and several hundred "cell pairs" of membranes and water flow spacers to achieve the required water production rate of the system. A single cell pair consists of four elements as shown in Figure 4-14.

- Cation selective membrane
- Ion depleting cell
- Anion selective membrane
- Ion concentrating cell

With multiple membranes and water passages located between a pair of electrodes it can be seen that alternating passages contain ion concentrating or ion depleting solution. The two flow paths are isolated and the ED process operates with a continuous flow of water

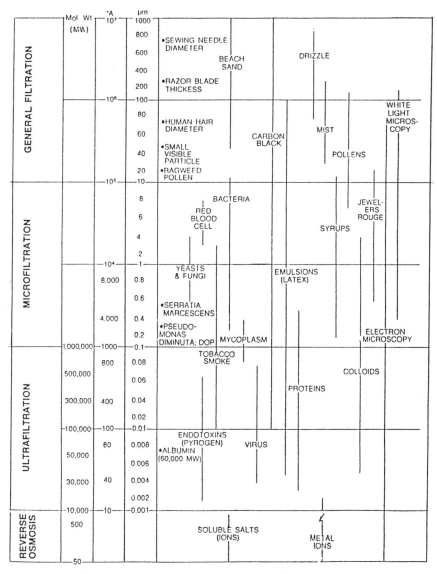

Å,ANGSTROM=10^{-8} cm   µm,MICROMETER (MICRON)=10^4 Å   1 mil=0.001 inch=25.4 pm
Differential pressure increases with reduced micron rate; dirt holding capacity and relative flow rates decrease with reduced micron ratings.

Figure 4-13. Relative Size of Small Particles

Source: Gelman Sciences Membrane & Device Division, Water Equipment Co., Ann Arbor, MI

through each cell. The reduction in concentration of ions in the depletion passage is a direct function of the amount of direct electric current applied and the flow rate of the water through the cell.

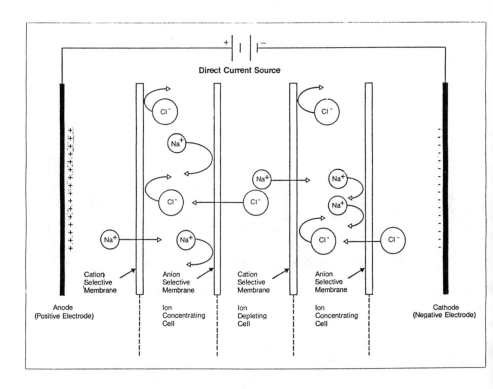

Figure 4-14. Operation of an Electrodialysis Cell

The electrodialysis membranes are designated as anion permeable (negative ion transfer), or cation permeable (positive ion transfer). Typical anions are: chloride, sulfate, nitrate, and bicarbonate. Cations include: sodium, calcium, and magnesium. The anion permeable membrane is an inert organic molecular structure having fixed positively charged groups in place in the membrane. Normally, this is a quaternary amine type of group. The structure of the cation permeable membrane is similar except that a fixed negative charge is in place such as a sulfonate group Under the driving force of an impressed direct current electrical field, positively charged cations will move in the direction of the negative electrode (cathode) until they are repelled by an anion membrane. Negatively charged anions will move toward the positive electrode (anode) until repelled by a cation membrane. Ions in solution of positive charge will electrically transfer through the cation membrane by means of the fixed negative charge groups. In like manner, negative ions will transfer through the anion membrane by means of the fixed positive charge groups. Negative ions in solution will be repelled by the negative fixed charges of the cation membrane with the opposite case being true for the anion membrane.

Water will exit from an electrodialysis cell, called a membrane stack, as two streams: a purified stream and a concentrate stream. In water desalting applications, the purified stream is the product while the concentrate goes to waste. The ratio of product to waste is normally controlled through concentrate recycle.

The removal of minerals from water by the electrodialysis process is a function of the amount of electric current applied. The process becomes more efficient with increasing temperature up to the present commercial limit of about 110°–160° F. Process efficiency is enhanced by hydraulic turbulence in the water passage areas to aid in the transfer of ions. Process inefficiencies can be caused by scaling or fouling of membrane surfaces. Membrane efficiency also depends on the construction and materials used to manufacture the membrane since there are many different types of membranes for differing applications.

Additive chemicals, such as acid, to the membrane stack may be required to avoid fouling of the membrane surfaces. Chemicals may also be added to improve stack operation and efficiency. This is accomplished by increasing the electrical conductivity of the water which reduces power requirements.

A special technique called polarity reversal has been developed to overcome membrane scaling and fouling problems without the addition of acids or other chemicals. This process reverses the electrical polarity of the electrodes three or four times per hour. At the same time, automatic valves reverse stream flow so that the depleting and concentrating flow paths are interchanged. This reversal causes ions to flow in the opposite direction breaking up scale forming concentrate layers at membrane surfaces. Colloids carrying electrical charges are also removed from the membrane and carried off in the waste stream. Most of the electrodialysis systems installed today use the polarity reversal system, and are generally referred to as Electrodialysis Reversal (EDR) systems.

EDR is unique among desalting processes in being capable of a reversal of this type because of the two inherent characteristics: a) EDR membranes are symmetrical—they operate the same way in both directions, and b) EDR stacks have a symmetrical configuration, with the concentrating and the diluting compartments being hydraulically and dimensionally identical.

Electrodialysis generally provides the most economic treatment for waters in the range of 1000 to 3500 ppm total dissolved solids (TDS). Typically, two and one-half kilowatt hours of electrical energy are required for each 1000 ppm of TDS removed from each 1000 gallons of product water. Pumping power adds about two and one-half kilowatt hours per 1000 gallons of product.

Electrodialysis should be considered where water contains between 500 ppm and 5000 ppm total dissolved solids (TDS). In the range of 500–2500 ppm, the ED system uses less energy than the reverse osmosis system. From 2500–5000 ppm total dissolved solids, ED may require more power than reverse osmosis.

Operation and maintenance costs vary with size of installation, number of stages, feedwater consumption, and product purity required. The projected cost per 1000 gallons for chemicals used to clean the stacks is 1 to 2 cents. Units use replaceable filter cartridges for final polishing, before the feedwater enters the ED plant. A rule of thumb cost for cartridge filters per 1000 gallons of product is

3 to 5 cents. Annual cost of membrane replacement depends on load factor, number of stages, and size of plant. Cost for replacement ranges from 10 to 20 cents per 1000 gallons of product.

Operating labor should average from one-half to one hour per day. Maintenance labor is more difficult to project, but generally 50 to 100 man-hours per stack per year is required, depending on the size of the plant and number of stages. Other expenses are the costs of the feedwater, wastewater disposal, building, pre- and post-treatment (if required), and amortization of capital investment.

Certain advantages are claimed for electrodialysis plants compared to reverse osmosis:

- Some manufacturers of EDR units state that, normally, chemical pretreatment is not required, as the reversal will clear any membrane deposition.

- ED plants operate at 30 to 60 psig, compared to a range of 200 to 450 psig to drive water through reverse osmosis membranes. Low pressure operation results in lower maintenance requirements and makes possible fabrication of wetted parts with non-corrosive plastics.

- Water from an ED plant is produced at nearly neutral pH. On the other hand, water from a reverse osmosis plant is normally acidic.

- In virtually all cases, ED gives the highest treated water yield (lowest blowdown).

- Presoftening is not required for ED.

- The membrane stack can be cleaned without damage to the system by polarity reversal, by flushing with an acid or caustic solution, or (in extreme cases) by opening stacks and scrubbing membranes. The ED stacks are the only membrane modules that can be cleaned in this manner. In contrast, reverse osmosis membrane modules must be discarded.

Reverse Osmosis (RO)

Reverse osmosis may be used to complete treatment for noncritical water supplies, or as pretreatment to reduce the cost of operating demineralizers.

Reverse osmosis (RO) is a process that separates salts as well as larger chemical components of a solution from water or other solvent by means of pressure (in excess of the "osmotic pressures") exerted on one side of a semipermeable membrane.

Normal osmosis takes place when water passes from a less concentrated solution to a more concentrated solution through a semipermeable membrane (Figure 4-15). A certain amount of potential energy exists between solutions on each side of the semipermeable membrane. Water will flow because of energy difference until the system is in equilibrium. The addition of pressure to the concentrated solution will stop transport of water across the membrane when applied pressure equals apparent osmotic pressure between the two solutions.

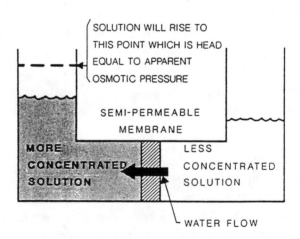

Figure 4-15. Osmosis

Source: Osmonics Inc.

As more pressure is applied to the more concentrated solution, water will flow from the more concentrated to the less concentrated solution. This process, known as reverse osmosis, is shown in Figure 4-16. The rate of water transport is a function of pressure applied to the concentrated solution, the apparent osmotic pressure (the difference in the absolute osmotic pressure of each solution), and the area of membrane.

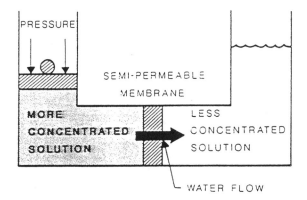

Figure 4-16. Reverse Osmosis

Source: Osmonics Inc.

Figure 4-17 is an artist's representation of the process of reverse osmosis. The solution being processed (the feed) is pumped under pressure across the membrane. Water in the feed solution is forced through pores in the membrane; that is, it permeates the membrane. Purified, or filtered, feed solution is called permeate.

It is important to remember that nearly every reverse osmosis plant includes a very efficient air stripping column for the finished water. The column is employed primarily for stripping carbon dioxide and is almost always a forced draft packed column. This means that volatiles that slip through the membranes are removed in the stripping column.

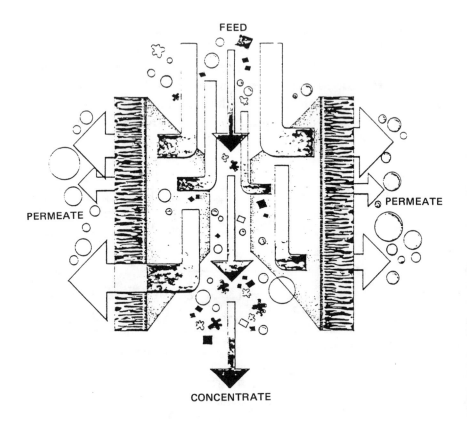

FEED

PERMEATE

PERMEATE

CONCENTRATE

Figure 4-17. Artist's Conception of Reverse Osmosis Process
Source: Osmonics Inc.

Depending on the characteristics of the ion solids in the raw water, the RO system will remove these with an efficiency of 70 percent to about 95 percent. The RO system is subject to precipitation, colloid coagulation, clogging by particulates, and microorganisms which can cause biodegradation of the membranes. Pretreatment of the water is therefore of the highest importance. The primary and usual-ly preferred pretreatment consists of filtration, various degrees of acid treatment with precipitation inhibitors, and finally, use of a bacteriocide, which is usually chlorine. Cellulose membranes will function with chlorine residuals up to 1 ppm; polyamide membranes will deteriorate in the presence of chlorine.

As a result of the removal of product water from the feed in the RO process, the residual brine concentration increases and the solubility limits of some solutes may be exceeded, causing precipitation. This is very much a problem with silica.

Because of the feed flows parallel to the membrane surface and only a portion of the flow permeates or diffuses through the membrane, the tendency for solids contained in the feedwater to deposit on the membrane is dramatically lessened compared with standard filtration processes. Flow velocities maintaining turbulence across the membrane act to limit membrane fouling. Nonetheless, depending upon the solids concentration in the feed-brine solution, some colloidal material does accumulate on the membrane surface, and in most instances, some form of membrane cleaning is required.

The typical reverse osmosis system requires filtration and pH adjustment of feedwater. In addition, performance may be limited by the concentration of constituents in the feedwater. Depending on operating conditions, membranes must be replaced on an average of every three to five years. These factors must be considered in cost estimates.

The energy required for the RO process is primarily for pumping. Typical energy consumption is 25–35 kWh per 1000 gallons of seawater. For brackish water, energy consumption ranges from 10–15 kWh per 1000 gallons. Power and labor are the largest components of the cost of operating an RO plant. Pumps are a critical maintenance item. For seawater, pumping pressure may approach 1000 psig; while for brackish water, pressure will be 300–400 psig. Overall, O&M costs for RO have been demonstrated to be lower than distillation or ion exchange. However, for situations where 12–15 ppm solids are objectionable, subsequent treatment by ion exchange is indicated.

Ion Exchange

Ion exchange is the most practical and flexible method of treating water for boilers and other processes. In the process, more desirable ions are exchanged for less desirable ones. The specific ions removed from solution by the ion exchange process are determined by the type of reactive material (resin) utilized and by the regenerant employed to provide exchangeable ions on the ion exchanger. The regenerant and regeneration process determine the ions that replace those removed from solution by the exchange process and provide the exchange capacity available to the ion exchanger.

Almost all dissolved inorganic material found in water solution is ionized, forming chemically equivalent amounts of cations and anions. Undesirable ions can be removed from solution and replaced with an equivalent amount of ions that are not objectionable by selection of the ion exchange process. If all dissolved, ionized material is objectionable, a combination of exchange processes operated in series can remove the ions and replace them with ions that combine to form water (hydrogen cations and hydroxyl anions). This process (demineralization or deionization) can produce water that is essentially free from dissolved ionized impurities.

There are some limitations to ion exchange. Silica is weakly ionized, and is difficult to detect either before or after the exchange. Colloidal silica, common in many areas, may foul the resin. In addition, chlorine, used for biological treatment of many water sources, is very deleterious to many resins, and should be removed prior to demineralization.

There are two fundamental types of ion exchange resins: cation resin, which removes some or all of the cations in the water (calcium, magnesium, and sodium), and anion resin, which removes some or all of the anions in the water (carbon dioxide, bicarbonate and carbonate alkalinity, chloride, and silica).

Cation resins fall into two general categories: strong acid and weak acid. Strong acid resins can remove all types of cations. Weak acid resins can remove only certain types of cations, but they operate at higher regeneration efficiency than strong acid resins.

Similarly, most anion resins are either strong base or weak base. Strong base resins can remove all types of anions, while weak base resins cannot. But weak base resins offer greater efficiency.

Strong base resins are further classified as Type I or Type II. The Type II resins are more efficient, but they lose the salt splitting characteristics at a lower temperature than the Type I resins. Type I resins, in addition to greater thermal stability, also provide better silica removal.

A further classification of resins is by structure—macroporous or gel. Gel resins are transparent beads with a low percentage of divinylbenzer "crosslinkage" in the polymer. Macroporous resins are opaque beads with a higher percentage of the crosslinkage. A final classification is bead size. The smallest resin beads tend to be washed away during backwash, and they tend to block the voids in the packed bed, increasing pressure drop. Larger beads also have their drawbacks— they have longer diffusion paths than smaller beads, so their capacity is not well utilized. Regeneration efficiency is also lower. Demineralizer designs should always specify the maximum and the minimum sizes of the resin beads.

Several ion exchange systems are listed in Table 4-6.

The most common form of ion exchange is water softening, where dissolved hardness is removed by exchanging calcium and magnesium ions for sodium ions. A strong-acid cation exchanger in the hydrogen cycle replaces all cations with hydrogen ions. The acidic water then passes through a strong-base anion exchanger in the hydroxyl (caustic) cycle that replaces all anions, including silica, with hydroxyl ions.

All ion exchangers with single beds of resin are essentially the same and vary only in the type of resin used, the regenerant chemical, and the materials of construction. A typical ion exchanger is shown in Figure 4-18.

The following description explains how single resin fixed bed ion exchangers operate.

Table 4-6. Ion Exchange Resin Systems

	CATION		ANION		
Resin	Weak Hydrogen	Strong Hydrogen	Strong Sodium	Weak Hydroxyl	Strong Hydroxyl	Strong Chloride
Removes	Calcium Magnesium (1)	Calcium Magnesium Sodium	Calcium Magnesium (2)	Chloride Sulfate (3)	Carbon-dioxide Bicarbonates Carbonate Chloride Sulfate Silica (3)	Bicarbonate Sulfate (4)

NOTES:
    (1) Hardness associated only with bicarbonate alkalinity
    (2) Water softener
    (3) Acid forms only
    (4) Dealkalizer

Courtesy of Drew Industrial Division, Ashland Chemical, Inc. Subsidiary of Ashland Oil, Inc.

Service: Water to be treated passes downward through the bed of ion exchange resin until its capacity for exchange is exhausted. Once exhausted, it can be regenerated.

Backwash: The flow of water is reversed and moves upward through the resin. This process removes accumulated suspended matter (resin is a good filter medium) and resin fines (broken resin beads). It also frees or fluffs the resin to insure good distribution.

Regenerant Introduction: Diluted chemical (salt, acid, or caustic) is introduced downward through the resin, removing exchanged ions, which are subsequently discharged. This restores exchange capacity.

Displacement: Water flows downward through the resin at a low flow rate, displacing (piston action) the regenerant chemical. This process is often called "slow rinse."

Rinse: A rapid downward flow of water removes all traces of regenerant chemical.

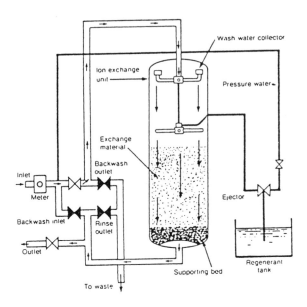

Figure 4-18. Typical Ion Exchanger Showing Surface Flow with Valve Nest for Regeneration-Backwashing, Injecting Chemicals, and Rinsing

Most ion exchangers are relatively complex systems that may experience malfunction. In troubleshooting ion exchanger malfunctions, accurate plant records of operation are important. Consideration should be given to the following:

Flows: Volume treated, normal flow rates, peak flow rate.

Pressure: Supply, and treated water, differential between units on multiple bed units; readings should be correlated with temperature and flow rate.

Resin: Type, bed depth, date of installation, results of any previous resin analyses. Consider possible fouling or temperature degradation.

Regeneration: Frequency, amount of concentrated chemical, time of introduction, concentration at resin, rate of flow of concentrated chemical and dilution water, source of dilution water, wash flow and duration, slow rinse flow and duration, fast rinse flow and duration.

Water: Analysis of raw water, treated water, and water between units on multiple bed units.

Design Criteria: Operating conditions for the system compared to original design criteria.

Ion Exchange Softening by Sodium Cycle

The most common form of single tank ion exchanger is the sodium cycle softener. A general configuration is to use two softener tanks piped in parallel, so that one tank is on stream while the other tank is regenerating, a procedure which may take several hours. Upon a preset signal—usually gallonage, but (more efficiently) a signal from a hardness monitor, the now exhausted tank starts to regenerate, and the second tank is put on the line. Depending on the ppm of TDS, the proper system must be selected. This is shown by Figure 4-19.

Capacity of a water softening resin is defined as a measure of the number of hardness ions removed from the water by a given amount of resin. This is usually expressed in grains per gallon (gpg). This capacity is affected by several variables.

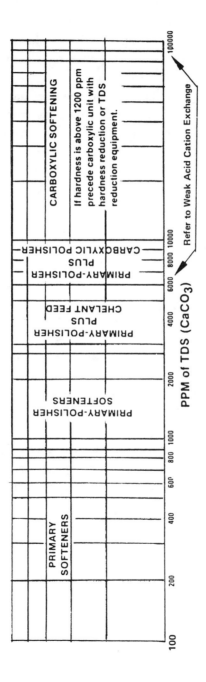

Figure 4-19. Selection of Softeners Based on TDS

Source: Struthers-Bruner, Winfield, KS

Resonance performance is affected by the type of action in the water. In addition to removing hardness (calcium and magnesium), salt regenerated softeners are capable of removing all positively charged cations in exchange for sodium.

Regeneration is the procedure that removes most of the hardness from the resin and leaves the resin in a state where it again can remove hardness ions from water. Salt, in the form of brine, serves as the regenerant because salt is relatively low in cost and contains a large number of sodium ions that are necessary to remove the hardness from the resin.

Increasing the amount of salt per regeneration increases the amount of hardness removed and thus increases the capacity. In addition to the amount of salt used, capacity is affected by the strength (concentration) of salt brine and by the contact time between the brine and the resin.

Salt Dosage: In conventional downflow regeneration, 15 pounds of salt will recover about 30,000 grains capacity, while 6 pounds of salt will recover about 20,000 grains. Thus, there is a great saving in total salt used by only removing the easy portion with 6 pounds of salt.

Regenerating at salt dosages below 5 pounds per cubic foot does not become more economical, but rather becomes inefficient, since the driving force of these small salt dosages is not enough to force the hardness out of the resin bed. Low salt levels also cause excessive hardness leakage. For this reason, softeners for oilfield steam generators are generally regenerated at dosages of 15 or 20 pounds of salt per cubic foot.

Brine Strength: Concentrated salt brine (100° Salometer, 2.6 pounds per gallon, 26 percent salt by weight) has a dehydrating effect on water-softening resin, causing it to shrink and making the hardness ions less accessible to the sodium in the salt brine. On the other hand, brine too dilute will not have the driving force necessary to regenerate the resin. In between these two extremes is a point where the brine is concentrated enough to affect regeneration

yet not so concentrated as to cause shrinkage. This point is 50°
Salometer (anywhere from 40° to 60° is acceptable).

When pumping brine from a brine pit or storage tank, it is important to dilute the brine to about 50° before it reaches the softener.

Brine Contact Time: Generally, a brine contact of 30 minutes,
with a maximum of 50 minutes, works best. At too high a flow rate,
brine passing through the resin doesn't allow time for efficient use of
the brine. Too slow a movement will allow equilibrium, and hardness
will pass back into the resin. Between these extremes is the optimum flow, slightly less than 1 gpm per cubic foot of 50° brine. This
combination of flow rate and contact time will normally be provided
by the eductor or brine-pumping-system.

TDS: For purposes of ion exchange it is essential that TDS be
reported as the ppm equivalent to calcium carbonate; to obtain the
calcium carbonate equivalent for TDS two methods are generally
acceptable and apply to almost all types of laboratory reporting. The
best method is to add the calcium carbonate equivalents of either
the cations (mainly sodium, calcium and magnesium) or the anions
(mainly alkalinity, chlorides, and sulfates). The next best method is
to take half of the specific conductance measured in micromhos as
the approximate ppm of TDS as calcium carbonate.

Flow Rate: As the rate of flow increases through the resin, hardness ions will be driven deeper into the bed before they are removed.
This causes an increase in the depth of the active exchange zone and
a decrease in capacity. The flow-rate effect on capacity is negligible
up to 6 gpm per cubic foot. At higher flows, the capacity is more
severely affected. On the other hand, extremely low flows of less
than 0.2 gpm per cubic foot will cause channeling, with resultant
severe and variable losses of capacity.

Bed Depth: The depth of the resin in the tank has an influence on
capacity. In a shallow bed, most of the depth is used by the active
exchange zone (e.g., a 20" bed with a 5" active exchange zone has
25 percent of the bed used up in the process of exchanging hardness).
Increasing the bed depth to 50", with the same 5" active zone,

would cause only 10 percent of the bed to be used by the active zone. Deeper beds have a greater overall capacity, since a greater percentage of the resin is available for softening.

End-Point Leakage: The term End-Point Leakage expresses the amount of hardness present in the effluent water at the point when a run is terminated and the softener is regenerated.

Leakage is defined as the continuous leakage of hardness ions occurring during the service run.

Leakage is caused by hardness ions left in the bed after regeneration. The hardness ions are bombarded by the sodium produced during the service run. These sodium ions have a regenerating effect on the remaining "left in" hardness, causing hardness to enter the soft water as hardness leakage.

At the beginning of the run, leakage is greatest because the sodium-rich soft water is "regenerating off" the hardness left at the very bottom of the bed. As the run progresses, this hardness at the bottom is eventually displaced with sodium. Therefore, the hardness leakage is greater at the beginning of the run and less just before final breakthrough.

Leakage can be reduced by using a higher salt dosage (i.e. leave fewer hardness ions in the resin at each regeneration) or by using a water with a lower TDS content (i.e. less sodium ions produced during ion exchange).

Leakage can also be reduced by regenerating the softener in an upflow direction. By passing a good grade of salt brine containing a very low calcium content upflow through the resin, the lower portion of the bed is highly regenerated and will contain very few hardness ions. During the succeeding downflow service run there will be fewer hardness ions in the lower part of the bed for the sodium to encounter.

Upflow regeneration will not be as efficient as downflow regeneration due to some expansion of the resin during regeneration but will result in lower hardness leakage.

Exhaustion: As hard water (containing calcium, magnesium, iron, etc.), passes through a bed of freshly regenerated resin, the hardness ions are removed in the upper portion of the bed. Depending upon several variables, the hardness ions may penetrate the bed a few inches or as much as several feet before they are removed. The depth of hardness ion penetration (i.e., vertical distance between the completely hard and the softened water) is known as the ACTIVE EXCHANGE ZONE. As hardness ions saturate the upper portion of the bed, the active exchange zone moves steadily downward throughout the run of the softener. Eventually the hardness ions reach the lowest point of the bed and hardness begins to appear in the outlet water.

The depth of the active exchange zone affects the capacity of the resin in the softener. Resin in the exchange zone is not available as capacity. Therefore, the more resin in the exchange zone, the lower the capacity of the resin bed.

Figure 4-20 illustrates the condition after regeneration.

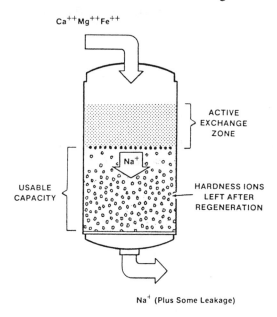

Figure 4-20

Primary Polisher-Softeners

Hardness leakage occurs because the sodium produced by ion exchange in the resin has a regenerating effect on any hardness remaining in the resin. By passing the water through a final polisher that has very little hardness remaining in the resin after regeneration, a more complete removal of hardness will result.

In a primary-polisher system the primary tank is regenerated downflow and will respond according to the previous discussion on primary softeners. The polisher installed in series following the primary unit is regenerated upflow. This leaves the lower part of the polisher bed almost free of hardness. The sodium in the soft service water passing downflow contacts little hardness in the resin and therefore leakage is very low.

The general configuration is to use two softener trains on a skid. Each train consists of a primary tank followed in series by a polisher tank. Only one train is on stream at a time. The other train is either regenerating or in the stand-by position (outlet valve closed). Upon a preset gallonage from a water meter in the outlet line of each train, the exhausted train starts to regenerate and the stand-by train is placed on stream (outlet valve open).

All four tanks are mounted on a steel skid with all interconnecting piping and wiring in place. Field installation requires only connection to inlet and outlet header, drain and electrical power. A briner may be located on the skid or salt brine may be made nearby and pumped to the softener train on demand. Figure 4-21 shows the general flow pattern through a single train.

To limit the leakage to 1 ppm, the TDS should not exceed 850 ppm with 20 pounds per cubic foot salting. This usage may be extended to about 2,000 ppm of TDS if a chelating agent is fed into the soft water to tie up the hardness leakage. Because chelating agents are expensive and are generally fed in ratios of 10 to 20 ppm for each ppm of hardness, the chelant method is limited to about 4 or 5 ppm of leakage. To limit the leakage to 1 ppm the TDS

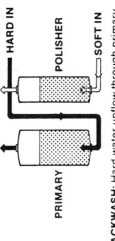

BACKWASH: Hard water upflow through primary. Soft water (from on-stream train) upflow through polisher. Time: 12 Minutes

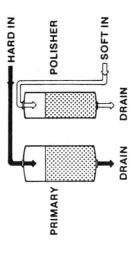

FAST FLUSH: Hard water downflow through primary. Soft water (from on-stream train) downflow through polisher. Time: 12 Minutes

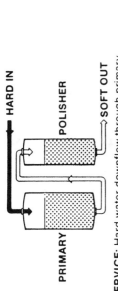

SERVICE: Hard water downflow through primary. Softened water downflow through polisher.

BRINE & SLOW RINSE: Brine pumped upflow through polisher and downflow through primary. Soft water slow rinse (from on-stream train). Time: 68 Minutes

Figure 4-21. Primary Polisher Flow Diagrams

Source: Struthers-Bruner, Winfield, KS

should not exceed 2,000 ppm for 15 pounds salting or 2,700 ppm for 20 pounds salting. This range may be extended to about 6,000 ppm if a chelating agent is fed with the soft water to tie up the additional hardness leakage.

An upflow regenerated primary softener will produce less overall leakage but capacity will also be less due to the general inefficiency of upflow regeneration.

Hardness must be in terms of total hardness as calcium carbonate. The hardness terms gpg (grains per gallon) is the same as part per million (ppm) or milligrams per liter mg/l divided by 17.1.

The softening resin is a strong acid type and regenerated with salt (NaCl) to the sodium cycle.

Flow rates through the resin should be no greater than 6 gallons per minute (gpm) per cubic foot in order to obtain maximum capacity.

Carboxylic (Weak Acid) Softeners

Unlike the strong acid resins used in softening, the weak acid resins are constructed of crosslinked polyacrylic acid and have weak acid (carboxylic) active exchange sites. Because the resin is so stable with acid, a regeneration with acid easily removes all of the collected hardness ions. The regeneration is very efficient, but the resin cannot be regenerated with brine. If operated on water with alkalinity less than the hardness, only the hardness equivalent to the alkalinity will be removed.

The resin has much greater capacity than salt regenerated softeners and capacities in excess of 40,000 grains per cubic foot are obtainable with some waters. Leakage is almost nonexistent with soft water effluent commonly in the range of 0.1 to 0.5 ppm hardness regardless of the TDS content of the water.

The weak acid resins are complete in their removal of hardness and effluents are all well below 1 ppm. At about 5,000 ppm of TDS the salt regenerated primary-polisher softeners are already leaking hardness in excess of 3 ppm and carboxylic softeners are frequently used following primary-polisher softeners to polish the hardness at TDS concentrations up to approximately 10,000 ppm. Over 10,000 ppm of TDS the salt-regenerated softeners are so inefficient from a capacity and leakage standpoint that carboxylic softeners become the exclusive ion exchange method. As the TDS increases above 10,000 ppm the carboxylic capacity decreases somewhat but will soften effectively up to a TDS of approximately 50,000 ppm.

The operating costs are affected largely by the hardness content of the water. Above hardness values of 1,000 ppm and TDS above 10,000 ppm it is well to investigate the operating costs of chemical lime soda softening and use a carboxylic softener to polish the effluent.

Each time the resin is regenerated, all the resin must first be acidified to the hydrogen form. Any hardness remaining will react with the caustic in the neutralization step to form a precipitate. It requires the same amount of acid regardless of the extent of exhaustion of the resin to the hardness form.

After acidification, an equivalent amount of caustic is used to convert the resin to the sodium form. Some economies are obtainable by reducing the caustic dosage with those waters having considerable alkalinity. The sodium cycle has the only disadvantage of the added cost of the caustic system.

Regenerating Sodium Exchange
Softeners with Potassium Chloride (KCl)

The perceived consumer concern about the excessive intake of sodium is an issue that has been widely discussed. Not only is excessive sodium intake a health hazard to many people, there are also indications that it can be deleterious to the general bodily welfare of the public. In many industries, such as food processing and in hospitals, the use of sodium regenerated softeners is frowned upon, as the softened water has an increased concentration of sodium compared to raw water. In such cases, the softeners may readily be regenerated using potassium chloride in lieu of sodium chloride. It is important to note that KCl will work in all cation exchange water softeners, and adds potassium bicarbonate to the softened water. As a water softener regenerant, KCl dissolves nicely and can reach nearly the same high level of concentration as does NaCl. The potassium ion has a greater affinity for the cation exchange resin than does sodium, so that the reaction will go very well. Experience has shown that most domestic water softeners do not require adjustment to operate effectively using KCl.

Another advantage in regenerating with KCl is that it is a primary nutrient for agricultural uses. It is known in agriculture as "potash," and is one of the three basic ingredients required for the healthy growth of plants. This may increase the options for discharging the regeneration stream. In some commercial or municipal applications the concentrated brine used in the regeneration stream could be stored separately and sold or disposed of as a fertilizer solution, therefore eliminating or reducing the disposal cost.

There need be no concern over the amount of potassium which may be ingested from drinking KCl treated water. As an example, assuming a daily fluid intake of approximately 2 liters per day, it could be said that on 20 grain water softened with KCl, it would add the same amount of potassium as the consumption of one banana.

Regeneration of Complex Ion Exchangers

The regeneration of multivessel ion exchangers is somewhat more complex than is the case with softeners. Figure 4-22 illustrates the steps in a typical regeneration cycle.

When regenerating a basic resin which has become exhausted to a high degree with calcium ions, regeneration with hydrochloric acid is recommended over the use of sulfuric acid to avoid calcium sulfate precipitation. However, economics may dictate the use of sulfuric acid as the regenerant. When sulfuric acid is used, the precipitation of calcium sulfate can be minimized or completely avoided by using stepwise sulfuric acid regeneration technique. In this technique, the heavily calcium laden resin is initially contacted with a 2 percent aqueous sulfuric acid regenerant followed by solutions of increased concentration, until the desired amount of acid has been consumed. Direct application of the usual 4 to 5 percent H_2SO_4 solutions for regenerating heavily calcium laden resins, may give rise to pressure drop problems during the subsequent exhaustion cycle resulting from the occlusion of $CaSO_4$ precipitate in the ion exchange bed, and calcium ions may appear as leakage during the service cycle. The result is a reduced working capacity and increased leakage, as well as excessive build-up of back pressure.

Alkalinity Reduction

In many cases, there is definite interest in removing only the bicarbonate or alkalinity that is present in water. This type of alkalinity is objectionable because the bicarbonates decompose under heat releasing carbon dioxide which causes corrosion problems. There are several ion exchange methods, which we will discuss below, that can handle this problem. However, the key to dealkalization by ion exchange lies in the ability of the hydrogen form of a cation exchange to convert bicarbonate salts to carbonic acid.

The most prominent of these methods are: sodium zeolite/hydrogen zeolite splitstream softening; sodium zeolite/chloride anion dealkalization; weakly acidic cation exchange; and avionic, although other systems are discussed.

Sodium Zeolite/Hydrogen Zeolite
Splitstream Softening

The sodium zeolite/hydrogen zeolite splitstream softening system consists of one or more brine regenerated sodium zeolite softeners and one or more hydrogen zeolite softeners containing strongly

1. SERVICE CYCLE

Water flows through the cation resin exchanging unwanted positively charged ions for hydrogen ions, and then through the anion resin exchanging unwanted negatively charged ions for hydroxide ions.

System remains in service until the conductivity sensing circuitry detects an increase of dissolved solids in the product water above the pre-set limit, indicating that the resins require regeneration.

The green GOOD QUALITY WATER indicator light is on.

2. SAFETY SENSE PERIOD

When poor quality water is sensed, the system attempts to rinse back to quality through the automatic drain. Automatic regeneration begins after the established safety sense period has elapsed.

The red POOR QUALITY WATER indicator light is on during the safety sense period and remains on during the regeneration cycle.

3. CATION BACKWASH CYCLE

Water flows up through the cation resin bed at a controlled rate, loosening the bed and preparing the resin for acid regeneration.

The amber CATION BACKWASH indicator light is on.

6. ANION BACKWASH CYCLE

Water flows up through the anion resin bed at a controlled rate, loosening the bed and preparing the resin for caustic regeneration.

The amber ANION BACKWASH indicator light is on.

7. CAUSTIC DRAW CYCLE

Sodium hydroxide passes through the anion resin, converting the resin back to the hydroxide form. Anions collected in the resin during the service cycle are rinsed to drain.

Caustic draw is terminated automatically when the pre-set time has elapsed.

The amber NaOH DRAW indicator light is on.

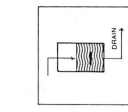
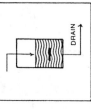

8. CAUSTIC RINSE CYCLE

This is a two-stage cycle: A slow rinse, in which spent caustic is rinsed to drain, is followed by a short fast rinse to prepare the resin bed for the next service run.

The amber NaOH RINSE indicator light is on.

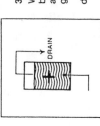

4. ACID DRAW CYCLE

Hydrochloric acid passes through the cation resin, converting the resin back to the hydrogen form. Cations collected on the resin during the service cycle are rinsed to drain.

Acid draw is terminated automatically when the pre-set time has elapsed.

The amber HCl DRAW indicator light is on.

5. ACID RINSE CYCLE

This is a two-stage cycle: A slow rinse, in which spent acid is rinsed to drain, is followed by a short fast rinse to prepare the resin bed for the next service run.

The amber HCl RINSE indicator light is on.

9. SAFETY FLUSH CYCLE

Water flows through the cation and anion resins, rinsing the system back to the pre-set quality point.

At the pre-set conductivity level, the automatic drain valve closes, the service valve opens, and service water is again available.

The amber SAFETY FLUSH indicator light is on.

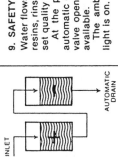

Figure 4-22. Typical Regeneration Cycle of an Automatic DI System

Source: Kane International, Inc., Rockford, IL

acidic cation exchange resin that are regenerated with acid. Water passes through the sodium zeolite and hydrogen zeolite units in parallel, and the unit effluents are blended to produce water of a desired alkalinity.

During a typical hydrogen zeolite service cycle, free mineral acids are produced throughout the service run with only minor variations prior to exhaustion. The treated water is almost completely free of calcium and magnesium ions. The amount of sodium leakage depends primarily on the ratio of sodium to total cations, the ratio of alkalinity to total anions in the raw water, and the resin regenerant dosage. In general, leakage increases as raw water sodium concentration increases, and decreases as raw water alkalinity and resin regenerant levels increase.

The concentration of free mineral acidity drops sharply as a hydrogen zeolite column nears exhaustion. At this point, the resin must be regenerated by treating it with an acid solution. Although other acids can be used, sulfuric acid is usually preferred for economic reasons.

The high acid concentration provides the driving force for displacing calcium, magnesium, and sodium ions, with hydrogen ions. As with the sodium zeolite softener, an excess of regenerant must be used.

The water produced by the sodium and hydrogen zeolite columns is soft (low in calcium and magnesium). Sodium zeolite water contains a mixture of sodium salts, and hydrogen zeolite water contains a mixture of acids. Sodium zeolite alkalinity (sodium bicarbonate) is used to neutralize the mineral acids.

In addition to neutralizing mineral acids, these processes convert sodium alkalinity to carbonic acid that is very unstable in aqueous solutions, and is, therefore, easily removed by degasification in a forced draft or vacuum degasifier.

Alkalinity of the treated water is maintained at the desired level by varying percentages of sodium and hydrogen zeolite water in the blend. For a given water, the higher the sodium zeolite percentage, the greater the alkalinity; the higher the hydrogen zeolite percentage, the lower the alkalinity.

The hydrogen zeolite/sodium zeolite splitstream softening system consists of one or more hydrogen zeolite columns, one or more sodium zeolite columns, a deaerator or degasifier, and a blending control system. Figure 4-23 is a flow diagram for a typical system.

The primary difference between the hydrogen and sodium systems is that the hydrogen zeolite system uses materials suitable for acid service.

The regenerant system is designed to handle sulfuric acid. A number of arrangements are provided for preparing the 2 to 6 percent acid concentration needed for regeneration. The most effective way uses in-line blending of concentrated sulfuric acid with dilution water that flows at a constant rate. Concentrated sulfuric acid enters the dilution line through a suitable mixing tee and the flow of concentrated acid is regulated to provide the desired regenerant concentration.

The blending control, which proportions the sodium to hydrogen zeolite flow, is usually a simple rate-of-flow controller, but automatic analyzers can be used to compensate for changes in raw water characteristics and to assure better control over the alkalinity of the treated water.

Degasifiers for removing carbon dioxide usually consist of plastic towers or packed columns with forced draft fans. The blended effluent from ion exchange columns is distributed over the top of the degasifier, and plastic slats or packing break the water into either fine droplets or films. Air, introduced at the bottom of the degasifier, flows countercurrent to the falling water and scrubs carbon dioxide from the effluent. Occasionally, vacuum degasifiers are used to remove carbon dioxide. A rubber-lined steel column is filled with packing to break the water into fine droplets and films, and steam jet ejectors, that apply a vacuum to the column, strip dissolved gases from the water.

Operation of hydrogen zeolite and sodium zeolite softeners is

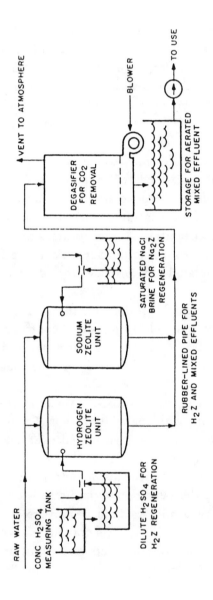

Figure 4-23. Hydrogen and Sodium Zeolite Units in Parallel

similar. Both use the same basic service and regeneration cycles. Service water and regenerant flows follow identical paths with the same flow rate restrictions. The only major differences are the methods for determining the end of the service cycle and the use of sulfuric acid rather than salt brine as the regenerant.

Sodium Zeolite/Chloride Anion Dealkalization

Sodium zeolite/chloride anion dealkalization systems have one or more sodium zeolite softeners and one or more chloride anion dealkalizers. Water first passes through the softener and then through the dealkalizer where bicarbonate, sulfate, and nitrate anions are replaced with chloride anions in the softened water.

The sodium zeolite/chloride anion dealkalizer system consists of one or more sodium zeolite softener columns and one or more chloride anion dealkalizer columns. As Figure 4-24 shows, the equipment is so arranged that the water flows in series from the sodium zeolite softener to the dealkalizer.

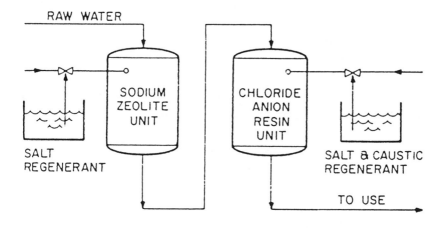

Figure 4-24. Sodium Zeolite/Chloride Anion Exchanger

The construction of the dealkalizer is nearly identical to that of the sodium zeolite softener. The major differences are the use of a strongly basic anion exchange resin in the dealkalizer and the provision for the addition of small amounts of caustic soda with the salt brine during regeneration.

The chloride anion dealkalizer differs from the sodium zeolite softener only in the specific controls used. The service water and regenerant flows follow the same routes through the equipment.

Weak Acid Cation Exchange

Another approach to softening and dealkalization uses weakly acidic (carboxylic) cation exchange resins. Water passes through a column containing this resin in the hydrogen form, and the calcium, magnesium, and sodium ions associated with alkalinity are exchanged for hydrogen ions. Carbonic acid is formed and a decarbonating tower may be added.

Raw water characteristics determine final treatment water quality. Because raw water cations associated with sulfate, chloride, and nitrate anions usually pass unaffected through the weakly acidic resin column, the treated water will contain hardness, if the raw water hardness exceeds the alkalinity. For example, raw water hardness and alkalinity levels of 150 and 100 ppm, respectively, produce a treated water hardness level of nearly 50 ppm. Consequently, the weakly acidic resin column may need to be followed by a sodium zeolite softener to remove the remaining hardness.

The regeneration frequency of a weak acid cation exchange resin column is usually determined by the treated water hardness or alkalinity level. Regeneration is accomplished by treating the resin bed with a sulfuric acid solution.

Since the regeneration efficiency of these resins approaches 90 percent, only slightly more than the theoretical quantity of acid is required to regenerate the columns.

The weak acid cation-exchange unit is identical to the standard hydrogen zeolite unit. The difference in performance is a result of using weakly acidic resin. The weakly acidic resin unit is followed by a sodium zeolite softener and a degasifier. Another arrangement provides a layer of strongly acidic cation resin beneath weakly acidic resin, rather than a separate sodium zeolite softener. Provisions for applying a salt wash after acid injection may be included.

Operation of a weak acid cation exchanger and sodium zeolite and hydrogen zeolite softeners is similar, differing only in specific controls used. Sulfuric acid is normally used for regeneration, but because weakly acidic resin displays a great affinity for calcium, the resin is more susceptible to calcium sulfate precipitation than strongly acidic resins used in the typical hydrogen zeolite unit. To avoid precipitating calcium sulfate, the user should closely follow equipment manufacturer's instructions.

Rohm & Haas Cationic Stratabed™ Neutralization: Another approach involving a weakly acidic cation exchange resin would be in the Stratabed configuration. In this case, the lower layered strongly acidic cation exchange resin operating in the sodium cycle. In conjunction with this resin the top layer would be a Stratabed grade weakly acidic cation exchange resin operating in the hydrogen cycle. In this particular case, the weakly acidic cation exchange resin would remove all of the hardness associated with the alkalinity. Any additional free mineral acidity that might be formed due to the weakly acidic resin would then be removed by the sodium form of the Stratabed grade strongly acidic cation resin. This resultant effluent would then be passed to a degasifier where all the carbon dioxide would be removed. (Figure 4-25)

High Alkalinity—Low Free Mineral Acidity (FMA) Waters: In waters of this type, another approach that can be used involves a strongly acidic resin operating in the hydrogen form. In this particular case, the effluent from such a system produces hydrochloric, sulfuric and carbonic acids. The effluent containing these acids would be sent to a degasifier to remove the carbon dioxide, and the resultant hydrochloric and sulfuric acids would then be neturalized with sodium hydroxide to give a neutral effluent. (Figure 4-26)

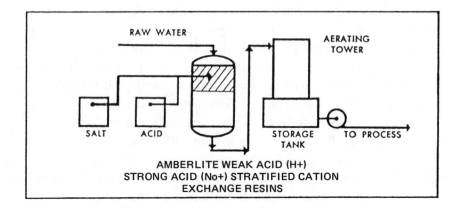

Figure 4-25. Dealkalization—Softening by Stratified Bed

Source: Rohm & Haas, Philadelphia, PA

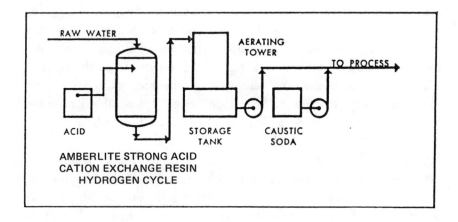

Figure 4-26. Neutralization by Caustic Soda

Source: Rohm & Haas, Philadelphia, PA

Raw Water Neutralization: Water of low hardness and high alkalinity can be treated, once again, with a strongly acidic cation exchange resin operating in the hydrogen cycle. In this particular case, the effluent from the cation exchange unit again contains hydrochloric, sulfuric and carbonic acids. This effluent can be neutralized with a side stream of raw water. The hydrochloric and sulfuric acids would react with the sodium bicarbonate present in the raw water forming sodium salts and liberating CO_2. (Figure 4-27)

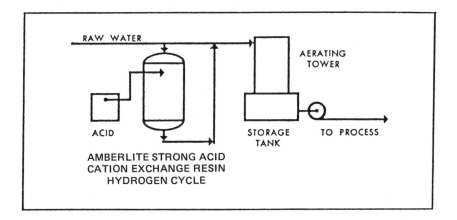

Figure 4-27. Neutralization by Raw Water

Source: Rohm & Haas, Philadelphia, PA

Anionic Dealkalization: Probably the most widely used system is one involving a strongly basic anion exchange resin. In this case, the resin is operated in the chloride form, and regenerated with sodium chloride. This is a relatively attractive approach, simply because one can use an inexpensive regenerant and eliminate an acid system with the sodium chloride regenerant, thus cutting down on the overall capital investment. In this particular case, all of the bicarbonates and sulfates would be picked up by the strongly basic anion exchange resin replacing them with the corresponding chloride salt. (Figure 4-28)

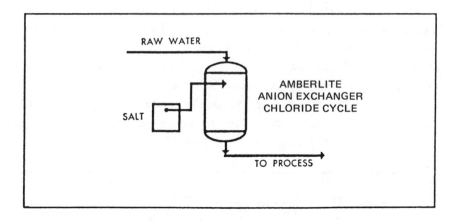

Figure 4-28. Dealkalization by Anion Exchange

Source: Rohm & Haas, Philadelphia, PA

Desilicizing

Softening and dealkalization processes do not reduce raw water silica content. Splitstream and weak acid cation exchange systems decrease dissolved solids to a limited extent, but chloride anion dealkalization does not significantly affect dissolved solids. Water produced by systems using forced draft degasifiers becomes saturated with oxygen, and consequently is highly corrosive.

For medium pressure boilers and low solids waters, removal of both the hardness and silica is necessary. In this case, one can use a strongly acidic cation exchange resin, operating in the sodium form, to remove the hardness, followed up by a strongly basic anion exchange resin. The strongly basic resin would be operated in the hydroxide form and regenerated with sodium hydroxide.

This type of system has the advantage of not needing any acid resistant equipment, and also can remove the silica, minimizing silica boiler problems. (Figure 4-29)

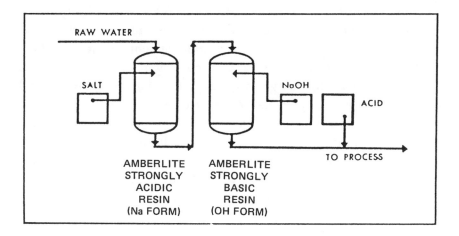

Figure 4-29. Desilicizing by Cation-Anion Exchange
Source: Rohm & Haas, Philadelphia, PA

Demineralization/Deionization*

Demineralization or deionization, as applied to water treatment, is removal of essentially inorganic salts by ion exchange. In this process, hydrogen cation exchange converts dissolved salts to their corresponding acids, and basic anion exchange removes these acids. The only other commercial process that produces water of comparable purity is distillation. Because demineralization of most fresh water supplies is substantially less expensive than distillation (evaporation), it is now more widely used.

A demineralizer system consists of one or more ion exchange resin columns, containing strongly acidic cation and anion exchange resins. The cation resin exchanges hydrogen for raw water cations and the anion resin exchanges hydroxyl anions for the highly ionized anions.

If the anion exchange resin is strongly basic, it will also remove such weakly ionized compounds as carbonic and silicitic acids. If the anion resin is weakly basic, it will not remove weakly ionized com-

*NOTE: Demineralization (DM) and Deionization (DI) are used interchangeably.

pounds. Because the exchanges are not 100 percent efficient, some mineral ions remain in the treated water. The presence of these ions is known as leakage. The amount of mineral contamination leakage in the demineralized water varies according to the demineralizer system used, and, for any given system, according to the raw mineral composition and demineralizer regenerant level (the amount of acid and caustic used for regeneration).

Demineralizers that are employed where silica removal is not critical use weak base anion resins, and specific conductance is the primary parameter for evaluating water quality. Such demineralizers use strong base anion resins, and both silica content and specific conductance are important water quality criteria. Both values are high after regeneration and low after rinsing to produce satisfactory water quality. The silica level, nearly constant during the entire service run, increases sharply at the end. Conductivity, also nearly constant during the service run, drops briefly at the end and then rises.

Demineralization systems for producing water of extremely high purity use mixed bed demineralizers (a mixture of strongly acidic cation and strongly basic anion resins in the same column).

Deionization Systems (DI) Operation

Deionizers operate like other ion exchange systems. The cation exchange resin columns operate in the same manner as hydrogen zeolite softeners. The anion exchange resin columns follow the same basic service and regeneration cycles. The only differences are use of an alkaline regenerant and specific controls. (Figure 4-30)

Ion exchangers are insoluble electrolytes, consisting of a high concentration of polar groups (acidic or basic) incorporated in a synthetic resinuous polymer vehicle. These resins are described as either cation exchangers or anion exchangers and the reactions of both types involve simple ionic equilibria. The following are typical reactions of cation exchangers, in which R^- represents the polymeric resin:

1. $R^-Na^+ + M^+X^- \rightleftharpoons R^-M^+ + Na^+X^-$
2. $R^-H^+ + M^+X^- \rightleftharpoons R^-M^+ + H^+X^-$
3. $R^-H^+ + M^+OH^- \rightleftharpoons R^-M^+ + H_2O$

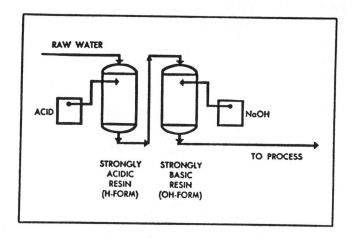

Figure 4-30. Deionization by Multiple Bed Exchangers

Cation Exchange: Equation 1 above represents the familiar water softening cycle where M+ is usually the calcium or magnesium ion. The reaction in the forward direction is the softening or water treatment step. In the reverse direction, this equation represents the regeneration of the resin with salt to remove the undesirable ions and reconvert the resin to the sodium form. The total ionic content of the water has not been changed, of course, but only an exchange of one cation for another has been effected.

Equation 2 is fundamentally the same as 1, the difference being that the hydrogen ion has been substituted for one of the metallic cations. This production of an acid from the corresponding salt is commonly called "salt-splitting" and is restricted to strong acid type cation exchangers. This equation is extremely important in deionization, since one of the products, i.e., free mineral acidity, can be removed by neutralization (see Equation 6 below).

Equation 3 illustrates the actual removal of an electrolyte by ion exchange, since the R^-M+ produced is completely insoluble to pure water.

Anion Exchange:

4. $R+Cl^- + M+X^- \rightleftharpoons R+X^- + M+Cl^-$

5. $R+OH^- + M+X^- \rightleftharpoons R+X^- + OH^-$

6. $R+OH^- + H+X^- \rightleftharpoons R+X^- + H_2O$

These three reactions of anion exchangers with $R+$ representing the resin, will be seen to be completely analogous to those in cation exchange. In particular, Equation 5 represents the production of an electrolyte which can be removed by the reaction of Equation 3 above and Equation 6 shows the removal of acidic electrolytes.

Deionization: The six equations shown above represent, as indicated, equilibrium reactions. The two neutralization reactions (3 and 6) tend to be only very slightly reversible, since one product, water, is ionized to only a very small degree.

Thus, if a salt solution were contacted first with a cation exchange resin in the hydrogen form, the resultant effluent, containing acids almost exclusively, could be stripped of electrolyte by contacting it with an anion exchanger in the basic form. Indeed, this scheme is the basis of the multiple-bed process of deionization shown in Figure 4-30.

This system is capable of producing water with a specific resistance of 250,000 to 1,000,000 ohms-cm. The total electrolyte content of water treated by this system can be as low as 2 parts per million. The fundamental reason for even this small amount of solids still remaining after the "two-bed" treatment is found in the fact that the "salt-splitting" reactions are completely reversible and some salt must inevitably leak through the bed of cation exchanger without having been converted to the corresponding acid. This is particularly true in the case of all economically feasible levels of regeneration. If the resin had been regenerated with a large excess of acid, this effect could be reduced further.

However, if the acid formed were removed as quickly as it was generated, the salt splitting reaction would then proceed to completion. Systems have been devised to accomplish this by employing alternate beds of cation and anion exchangers involving as many as six or more columns in series.

Another solution to this problem of promoting unfavorable equilibria is to mix the cation and anion exchange resins intimately, thereby insuring that the acid formed by contact of a salt initially with a particle of cation exchanger would be immediately neutralized by the neighboring particles of anion exchanger. The alakli formed by initial contact of the salt with the base anion exchanger would also be removed immediately by the cation exchange resin. This arrangement is referred to as a "mixed bed" or "monobed" unit. This system would then be capable of "pushing" even very unfavorable equilibria to completion. Even for substances which are only slightly dissociated, such as silicic acid, this method could effect virtually complete removal by the simultaneity of the reactions involved if strongly acidic and strongly basic resins were used. Some silica present in colloidal form may, however, even leak through such resin mixtures.

In the mixed bed ion exchanger, the anions and cations are mixed together in the exchange tank, providing, in effect, an infinite series of anion/cation sets. The effluent from a mixed bed unit does approach theoretical purity, limited only by the contamination resulting from the materials which contain the exchange process. Regeneration of the mixed bed exchanger requires that the anion and the cation resins be separated, a procedure referred to as "classifying." This is possible because the anions have a different specific gravity from the cations.

Mixed Bed System Regeneration (Figure 4-31)

Hydraulic Separation (Backwash): In order to separate the resins into two discrete zones for regeneration, water is introduced upflow through the bottom of the unit. This operation is commonly known as backwashing.

Two physical characteristics are important in determining the hydraulic suitability of materials for the Monobed resin technique. First, the resins must be of sufficiently different density to make the separation clean-cut. Resins which do not separate well in water because of too small a difference in buoyancy can often be fractionated by employing a brine of intermediate density.

Second, the resins must have the proper particle size distribution to insure good separability. The various resins are, of course, carefully checked for the proper particle size distribution and are "tailor-made" for the most efficient operation.

Following the backwash operation, the resins are allowed to settle into two distinct zones after which they are ready for regeneration.

Regeneration: Two basic methods are used for the regeneration of the separated bed.

Two-Step Method: In this scheme, the anion exchanger is regenerated first, with alkali flowing from the top distributor through both the anion and cation exchangers and out through the bottom of the column. Alternatively, the alkali can be taken off at the center of the bed by filling the voids in the cation exchanged bed with water.

The cation exchanger may be regenerated either upflow by introducing the acid through the bottom and out by way of the central distributor located at the resin boundary or, alternatively, may be regenerated downflow from the central distributor.

Simultaneous Method: In this method, both resins are regenerated at the same time. Acid is passed upflow through the cation exchanger and alkali is passed downflow through the anion exchanger. Both spent regenerants are withdrawn through the central distributor at the resin boundary.

In general, it is safe to say that the Monobed resin operation may require a somewhat more complex regenerant system than a two-bed installation. It is equally safe to say, however, that

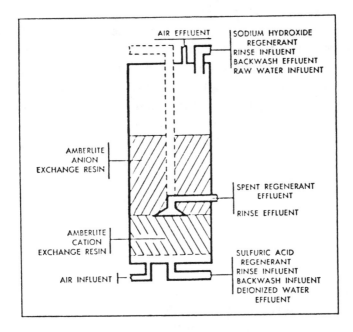

**Figure 4-31. Typical Arrangement of a Monobed
Ion Exchange Resin Deionization Unit**

Source: Rohm & Haas Company, Philadelphia, PA

the quality of effluent from that operation, attainable by no other commercial method, is of itself more than ample justification for its use.

The type of regenerant system most satisfactory in a particular instance will depend on several considerations. The advice of the Rohm & Haas Company or other resin manufacturers' ion exchange technicians and competent water conditioning equipment manufacturers should be sought in all instances.

Rinsing: Both the anion and cation exchangers must be rinsed after regeneration. In both the two-step and simultaneous methods, the rinses should follow the same path as the regenerant solutions to avoid exhausting either resin after they have been regenerated. The rinse is continued until the bulk of the regenerant has been eliminated but it is not necessary to remove the last traces. The small amounts of remaining regenerant can easily be eliminated by finishing the rinse after the resins have been mixed. This final rinse promotes complete removal of all traces of regenerant by ion exchange, at the expense of a small amount of resin capacity. This technique may reduce rinse requirements by one-half.

Mixing the Resins: After the resins have been rinsed, they are mixed by means of compressed air injected from the bottom of the column. The mixing must be complete to insure highest quality effluent. Unmixed zones, particularly at the bottom of the bed, may lead to hydrolysis with consequent electrolyte leakage. It has been found that, if the water above the resin is drained to within an inch or two of the top of the bed, maximum turbulence is obtained during the air mixing operation.

Because of the difficulty of classifying the resins in the exchange tank, a buffer resin is sometimes provided, which has a specific gravity such that it stratifies between the anion and the cation layers. Figure 4-32 explains the use of buffer beads.

Other Processes

Recent developments in ion exchange have led to the use of new techniques for handling problems and to more economical demineralization approaches. One relatively new development has been the commercial application of continuous, countercurrent ion exchange which uses ion exchange resins more efficiently than the traditional fixed-bed systems. This process passes the water through the resin column so that it first penetrates nearly exhausted resin and progressively reaches resin in higher stages of regeneration until, just before leaving the column, it passes through freshly regenerated resin. Exhausted resin is frequently removed from the service vessel and regenerated in separate units and then returned to the main vessel. Backwashing is required only after 20 to 40 cycles. This technique results in very low leakage.

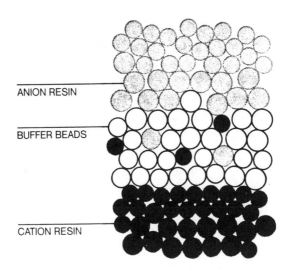

ANION RESIN

BUFFER BEADS

CATION RESIN

In a mixed bed system, cation and anion resin must be separated before regeneration. Since cation resin is more dense than anion resin, backwashing leaves most of the cation resin at the bottom of the bed. But undersize cation beads may end up among the anion resin—and oversize anion beads may settle among the cation resin. These misplaced beads are then "regenerated" with the wrong regenerant, and this cross-contamination leads to leakage and capacity loss.

One way to effectively improve the separation of cation and anion resin in a mixed bed is with Buffer Beads. This new inert resin forms an intermediate density layer between carefully selected cation and anion resin. While this doesn't change the fluidization properties of cation and anion beads, it does dilute the area of crossover. So far fewer resin beads will be out of place during regeneration. Plus, the white Buffer Beads provide a clear-cut distinction between anion and cation layers that's much easier to see.

Figure 4-32. Buffer Beads Improve Separation in Mixed Bed Systems

Another significant development uses layered resin beds, usually weakly acidic and strongly basic anion resins, in the same column. Because the weakly basic resin is less dense than the strongly basic resin, backwashing after regeneration causes the weak base resin to form a layer on top of the strong base resin. Figure 4-25 is an example of a layered bed.

If the weak base resin is selected properly, the layer it forms will protect the strong base resin from organic fouling and will also economically remove mineral acidity. This can also be applied to cation units where weakly acidic (carboxylic) cation resin is used in the same vessel as strongly acidic cation resin. In both cases, significant regenerant savings can result.

Limitations

Like other ion exchange systems, demineralizers will function efficiently only if the water supply is free from suspended matter and oxidizing materials such as chlorine. Anion exchange resins are more susceptible to oxidants than the cation resins, thus the removal of oxidants is necessary for good demineralizer performance over an extended period of time. Standard demineralizer resins cannot remove colloidal or some organically sequestered materials and their unexpected presence has caused silica or iron contamination in some demineralized water supplies. While some organic materials will pass through demineralizer columns unaffected, others may seriously foul the anion resin beds.

Equipment

The demineralization system varies according to requirements of the application. Figure 4-33 shows nine basic demineralizer systems and lists areas of application, typical treated water quality, and the relative advantages and disadvantages of each.

The cation and anion exchange columns used for demineralization are similar to those described for hydrogen zeolite softening. The only difference is the use of anion resin in the anion exchange column. Piping, valves and fittings are of either lined or acid resistant materials. When required, either the deaerator or the vacuum degasifier is identical in design to the unit used in split-stream softening.

The mixed bed demineralizer is similar to the hydrogen zeolite softener, except that the resin bed is a mixture of cation and anion exchange resins.

Combining Reverse Osmosis (RO) and Dionization (DI) Systems

For many years there was essentially only one economical way to demineralize water, regardless of total dissolved solids (TDS): ion exchange. But since ion exchange (IX) is an inherently quantitative process, the higher the influent TDS, the faster the available exchange sites become filled and the more often the resin has to be regenerated. (We recognize that flow rate and desired water quality have an effect, but they are normally fixed by the process design.) Operating costs then, are directly related to influent TDS (usually below 1000 ppm), and there is no way to avoid it.

Reverse osmosis (RO), on the other hand, is a more qualitative process, removing 90+ percent of the TDS in the influent. RO is a likely candidate for brackish water treatment, and, in fact, has been . . . for almost two centuries. It has only been in recent years, however, that RO made the critical transition from the laboratory to a field-proven, economically attractive water demineralization method, largely due to advances in membrane technology.

Although the combination RO/IX approach discussed here is not universally applicable, in situations where it can be used it will yield impressive economies.

As mentioned earlier, flow rate and water quality are fixed in most systems, and the chemical operating costs of an ion exchange system are a function of influent TDS. Any reduction in TDS will give a reduction in chemical costs. By using RO as a roughing demineralizer ahead of an ion exchange system, the user will remove 90+ percent of the influent TDS and reduce both regenerant and chemical usage by about the same amount. And since less regenerant is used, less appears in the waste stream, an added environmental and cost bonus.

DEMINERALIZER SYSTEM	APPLICATION	TYPICAL EFFLUENT	ADVANTAGES AND DISADVANTAGES
	Silica and CO_2 are not objectionable.	Specific conductance 10-30 micromhos. Silica unchanged.	Low equipment and regenerant costs.
	Silica is not objectionable but CO_2 removal is required.	Specific conductance 10-20 micromhos. Silica unchanged.	Low regenerant costs, but requires repumping.
	Low alkalinity raw water, silica removal required.	Specific conductance 5-15 micromhos. Silica 0.02 to 0.10 ppm.	Low equipment costs, repumping not required, high chemical costs.
	High alkalinity, raw water, silica removal required.	Specific conductance 5-15 micromhos. Silica 0.02 to 0.10 ppm.	Low chemical costs, repumping is required.
	High alkalinity, sulfate and chloride raw water, Silica removal required.	Specific conductance 5-15 micromhos. Silica 0.02 to 0.10 ppm.	Low chemical costs, high equipment costs, repumping required.

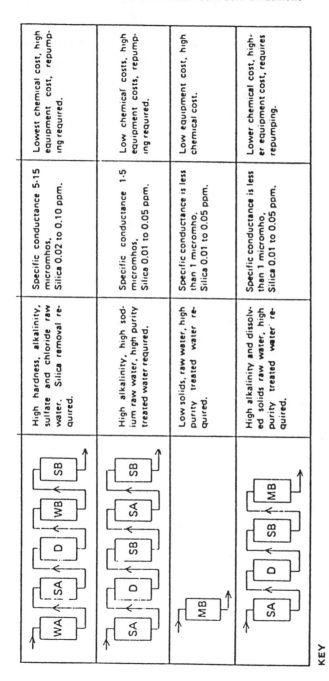

Diagram	Raw water condition	Effluent quality	Cost notes
WA → SA → D → WB → SB	High hardness, alkalinity, sulfate and chloride raw water. Silica removal required.	Specific conductance 5-15 micromhos. Silica 0.02 to 0.10 ppm.	Lowest chemical cost, high equipment cost, repumping required.
SA → D → SB → SA → SB	High alkalinity, high sodium raw water, high purity treated water required.	Specific conductance 1-5 micromhos. Silica 0.01 to 0.05 ppm.	Low chemical costs, high equipment costs, repumping required.
MB	Low solids, raw water, high purity treated water required.	Specific conductance is less than 1 micromho. Silica 0.01 to 0.05 ppm.	Low equipment cost, high chemical cost.
SA → D → SB → MB	High alkalinity and dissolved solids raw water, high purity treated water required.	Specific conductance is less than 1 micromho. Silica 0.01 to 0.05 ppm.	Lower chemical cost, higher equipment cost, requires repumping.

KEY

| SA | STRONGLY ACIDIC HYDROGEN CATION | WA | WEAKLY ACIDIC HYDROGEN CATION | WB | WEAK BASE ANION | SB | STRONG BASE ANION | D | DEGASIFIER OR VACUUM DEAERATOR | MB | MIXED BED |

Figure 4-33

Source: Betz Laboratories

This 90 percent reduction in influent TDS considerably reduces the load on the ion exchange resin beds, providing an approximate tenfold increase in run length. The need for standby demineralizers is lessened and overall ion exchange system size may be much smaller. Figure 4-34 shows the set up of a 2-bed vs. a 3-bed IX system.

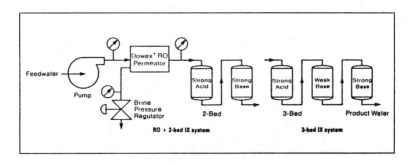

Figure 4-34. Reverse Osmosis + 2-Bed Ion Exchange System vs. 3-Bed Ion Exchange Systems
Courtesy of Dow Chemical Co.

Figure 35-A graphically illustrates the operating costs of both systems as a function of the TDS in the water going to the system involved. Note that the RO + IX curve increases very gradually with TDS, while the ion exchange curve has a much steeper slope. The crossover point of 375 ppm would be for a new system.

Another way to view the economics of these two approaches to demineralization is provided in Figure 4-35B where the savings in regenerant and other chemical costs for an RO-equipped ion exchange system are related to the reverse osmosis system operating costs. RO costs include membrane replacement at 3-year intervals, power to run the system and chemical costs. While some RO systems require no chemicals, in practice chemicals are usually required to control pH, iron fouling, scale, and biological growth.

Figure 4-35B also clearly shows that when TDS exceeds 320 ppm, the savings in operation costs for the ion exchange system pay for the cost of operating the reverse osmosis system. Above this TDS "break-even" point, the dollar difference increases in direct proportion to the TDS in the influent, and the user enjoys yet larger net gains.

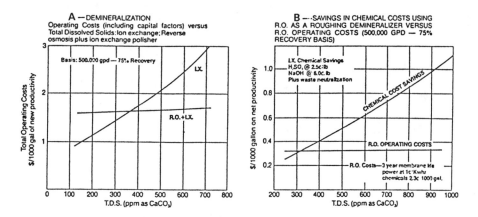

Figure 4-35
Courtesy of Dow Chemical Company, March 1991

Table 4-7 shows the sensitivity that these curves have to changes in the base condition and may be used to estimate the economic equality point between a 3-bed ion exchange demineralization system using countercurrent cation regeneration and one using an RO roughing demineralizer preceding a 2-bed ion exchange system.

Combining Electrodialysis Reversal (EDR) and Deionization (DI) Systems

Water quality attainable with EDR is primarily a function of: 1) the membrane stack area; 2) applied d-c voltage; and 3) the total dissolved solids in the water.

In a similar fashion water purity attainable through ion exchange is a function of: 1) resin bed volume; 2) chemical charge; 3) flow rate; and 4) the total dissolved solids in the water.

Table 4-7. Calculation of TDS at Equal Operating Costs for Reverse Osmosis and 2-Bed Ion Exchange vs. 3-Bed Ion Exchange

	(375 ppm T.D.S. FOR BASE CASE)		
PARAMETER	BASE CASE CONDITION	NEW CONDITION	CHANGE IN T.D.S. (PPM AS $CaCO_3$)
PRODUCTIVITY	500,000 gpd	100,000 gpd	+85
		300,000 gpd	+30
		700,000 gpd	−25
		900,000 gpd	−45
		1M + gpd	−50
CHEMICAL COSTS	NaOH at 6¢/lb.	NaOH at 9¢/lb.	−25
		NaOH at 12¢/lb.	−50
	H_2SO_4 at 2.5¢/lb.	H_2SO_4 at 3¢/lb.	−25
WATER RECOVERY WITH R.O.	75% RECOVERY	85% RECOVERY	−20
		90% RECOVERY	−25
CAPITALIZATION FACTOR	(0.1565) D.F.C. (UTILITY INDUSTRY PRACTICE)	(0.3) D.F.C. (CHEMICAL INDUSTRY PRACTICE)	+55
MEMBRANE COSTS	20¢/gpd	18¢/gpd	−16
MEMBRANE LIFE	3 year life	4 year life	−25
WATER DISPOSAL COSTS	$0.10/1000 gal. WASTE WATER	ADD $0.01/lb. of T.D.S.	−65
		ADD $0.02/lb. of T.D.S.	−110
POWER COSTS	$0.01/Kw. Hr.	ADDITIONAL $0.01/Kw. Hr.	+25

Courtesy of Dow Chemical Company

Both systems remove solids by charge transfer. The EDR is able to continuously discharge ions whereas the DI must intermittently be recharged or regenerated.

Consequently, it is intuitively obvious that utilization of EDR to pretreat ion exchange feedwater will increase yield and purity from the DI with a reduced charge depletion of the resin bed.

A number of installations have shown that pretreating DI feedwater with an EDR will significantly reduce the necessity for regeneration requirement of the DI unit. In addition, the quality of water from a two-bed system can go from 200,000 ohm maximum to well over a megohm. Return on capital investment is usually excellent, with paybacks in the order of one year. The increased capacity of the DI units, plus better quality water, may result in significant annual savings.

Recent Advances in Ion Exchange Technology

New methods of applying ion exchange technology are available. Although they will not immediately replace current methods, they are useful for specific purposes, and should be considered on the basis of cost, and applicability to special requirements.

The Ionpure System

This is a continuous deionization system which requires no regenerant chemicals, does not become exhausted, and produces 1 megohm-cm water. Figure 4-36 explains the principle of electrodeionization.

The Recoflo System (Figure 4-37)

Description: Recoflo water deionization systems follow the basic ion exchange principle, whereby ions in solution are exchanged for other ions contained inside exchange resins. But here the similarity to conventional systems ends.

In any operating ion-exchange column, the exchange resin near the top is exhausted and the resin near the bottom is in the regenerated state. At any given time, most of the resin is inactive, with ion exchange taking place only within a very narrow band of resin called the exchange zone. This exchange zone moves down the column unit until all the resin is exhausted.

Recoflo systems utilize resin columns only slightly longer than the exchange zone—in most cases, only a few inches high.

The following design features are incorporated to make the short columns work:

Fine-mesh resins provide a larger number of readily accessible exchange sites near the resin surface where ion exchange takes place most quickly. Smaller resin particles also have greater physical stability and can be rinsed more efficiently.

Low exchanger loading utilizes only those ion-exchange sites near the resin-bed surface—the most accessible sites.

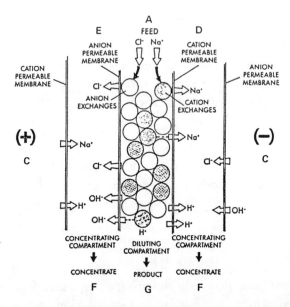

A) Feedwater enters the system and flows inside resin/membrane compartments. Part of the feed stream flows along the outside of the membrane surface to wash away unwanted ions.
B) Resins capture dissolved ions.
C) Electric current drives captured cations through cation membranes and captured anions through anion membranes.
D) Cation-permeable membranes transport cations out of resin compartment, but prevent anions from leaving waste stream.
E) Anion-permeable membranes transport anions out of resin compartment, but prevent cations from leaving waste stream.
F) Waste steam flushes concentrated ions from system.
G) Product water leaves system.

Figure 4-36

Source: Millipore Corporation

Counter-current regeneration improves loading and regeneration efficiency.

No column freeboard reduces liquid dilution, intermixing, and resin attrition by eliminating the free space in the column above the resin.

Short cycle times. High flow rates, low loading and small resin columns allow regeneration to be completed in a few minutes.

Details of the deionizer operating cycles are as follows:

Onstream: Prefiltered water is pumped first through the cation column to exchange cations in water for hydrogen ions (H+), then through the anion column to exchange anions for hydroxyl ions (OH−). Deionized water then leaves the anion column for use as required in the process.

The unit remains onstream until the exchange resins approach exhaustion. This point can be determined in a variety of ways—monitoring of deionized water conductivity, or simply by cycle timing. The onstream cycle is automatic and varies in duration between 5 and 60 minutes, depending on raw water composition.

Regeneration: Dilute acid and caustic regenerant are pumped countercurrently through the cation and anion columns respectively.

Regeneration rinse: Water flows up through the resin beds to rinse out the regenerant chemicals. The regeneration and rinse steps generally take less than two minutes.

Recirculation: Water contained within the system's plumbing and resin beds is circulated through the cation and anion beds until the water is of the desired quality (by conductivity).

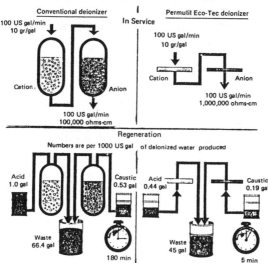

Figure 4-37. A Comparison Between Recoflo and Conventional Deionizer Performance
Source: ECO-TECH LTD., Ontario, Canada

CHLORINE IN HIGH-QUALITY WATER SYSTEMS

Chlorine can reduce the capacity of ion exchange systems and shorten the life of the resins. It has the potential to attack reverse osmosis membranes, and can cause chloride stress corrosion in piping. In some cases, however, it is necessary to control the growth of some types of biological entities. This is evident in its use in conjunction with activated carbon filters, for instance.

When chlorine gas dissolves in water, it hydrolyzes rapidly according to the following equations:

1. $Cl_2 + H_2O \rightleftharpoons H+ + Cl^- + HOCl$

2. $HOCl + H_2O \rightleftharpoons H_3O+ + OCl^-$

The oxidizing or sterilizing agents are hypochlorous acid (HOCl) and hypochlorite (OCl). However, the extent to which each is present depends on the pH of the water. At pH values above 3.0 and chloride ion (Cl^-) concentrations below 1,000 ppm, free chlorine (Cl_2) molecules are virtually nonexistent. The ratio of hypochlorous acid to hypochlorite. Equation 2 is totally dependent upon the pH of the water and its temperature as shown in Table 4-8.

The data shows that below pH 7.5 hypochlorous acid is predominant. Above pH 7.5 hypochlorite is predominant. It should also be noted that chlorine hydrolysis reduces alkalinity. In weakly buffered waters, it may cause a measurable pH decrease due to the formation of hydrochloric acid (Equation 1).

The fact that analytical tests for "chlorine" report results as "free residual Cl_2" or "free available Cl_2" may lead people to believe that chlorine is present as a gas. As we can see from Figure 10-16 and Table 4-8, this is not so. Above pH 3.0 chlorine is not present in water as a gas. Above pH 3.0 "free residual chlorine" is a measure of hypochlorous acid, and "free available chlorine" is a measure of hypochlorous acid, hypochlorite and chloroamines.

The pH dependent relationship of chlorine with water is important, because hypochlorous acid (HOCl) is a better antibacterial agent than hypochlorite (OCl), which is predominant over a pH of about 7.5. OCl cannot diffuse through the cell walls of microorganisms. This explains why the disinfecting efficiency of "free residual chlorine" decreases as pH rises. It is almost totally ineffective above a pH of 9.

Table 4-8. Hypochlorous Acid Concentration as a Function of pH and Temperature

	%HOCl	
pH	$0°C$	$20°C$
4	100	100
5	100	99.7
6	98.2	96.8
7	83.3	75.2
7.5	61.26	48.93
8	32.2	23.2
9	4.5	2.9
10	0.5	0.3
11	0.05	0.03

Use with DI Units: Any level of chlorine is potentially harmful to ion exchange resins. The maximum permissible chlorine residual for water entering an ion exchange demineralizer or softener is generally considered to be 0.5 ppm. Therefore, if the water supply is municipal, excess chlorine should be removed before the water enters the treatment system.

Municipal water is treated for the residential user, and municipalities will not usually alter their treatment techniques for an industrial user, no matter how large. Therefore, it is essential to establish communications with your water supplier so that you will be made aware of any process changes that may affect your operation.

It has been widely believed that residual chlorine is removed in the cation resin bed and never reaches the anion resin. There is growing evidence that this is not true. A number of installations have measured residual chlorine after the anion resin bed and even after mixed bed polishers.

Effect on RO Membranes: Many water treatment systems employ reverse osmosis. The most common reverse osmosis membranes are spiral-wound cellulose acetate and hollow-fiber polyamide. The impact of chlorine on these membranes is critical.

Manufacturers of cellulose acetate membranes recommend a maximum free residual chlorine of 0.5 ppm throughout the system to maintain a low level of bacteria growth. They also recommend a water temperature of 70 to 75°F and a pH of 5.0 to 6.0. Thus most of the residual chlorine is present under these conditions as hypochlorous acid, the more effective sterilizing agent.

Manufacturers of polyamide membranes recommend removal of all residual chlorine before the water enters the reverse osmosis unit. Apparently, chlorine affects the life and operation of this type of membrane.

USING ION EXCHANGE TO REMOVE
DISSOLVED AND COLLOIDAL SILICA

Small amounts of silica occur in most natural waters, and larger amounts in waters which originate in volcanic-type terrain. In the range of 5-20 ppm, silica is not classed as a major component, and in fact is sometimes added to water to assist in coagulation and clarification. When the water is used to feed high pressure boilers, over 400 psig, it becomes a serious contaminant due to its volatilization in steam. High pressure steam turbine plants find volatilized silica to be a serious nuisance, as the vapor condenses on the blades of the turbines, causing unbalance, erosion and loss of efficiency.

Silica is now commonly removed by ion exchange; there is, however, a problem of detecting the end point of an operating run, and the possibility of considerable leakage toward the end of the run. Since silica is very weakly ionized, conductivity measurements for resin exhaustion are not very reliable. Fortunately, there are now sophisticated silica analysis systems which can detect silica in the ppb range. The problem of leakage is mitigated to a large extent by the use of heated caustic for regeneration of the anion bed. A limitation here is the resistance of the resin to temperature degradation. Using currently available designs and techniques, silica level in effluent water can be reduced to 1.0 ppb.

Silica has a complex chemistry with the ability to form polymers or gel-like compounds depending on conditions. The analysis of silica in water has been to a large extent performed by the molybdate method with considerable accuracy to about 5 ppb. However, the method works only with dimers, so higher polymers go undetected. Silica values not measured by the molybdate method are classed as colloidal silica, and may represent trimers and other low molecular weight polymers, and range on up to quite large particles easily filtered by absolute filters. When in the range of several ppb, determination may be made gravimetrically using a total weight loss on fluoric acid digestion and evaporation. This difference will be total silica, and consists of dissolved (detected by the molybdate method) plus the colloidal component.

When purchasing a silica removal deionizer, it is always important to determine beforehand the amount of colloidal silica, and its nature, so the proper resins may be selected. Then, when the system has become operational, attention should be given to the effluent with respect not only to dissolved, but to colloidal silica.

In the design of the system, the selection of Type I or Type II resins will be determined by the method of regeneration. Further, in recent usage, it has been found that countercurrent regeneration designs usually result in the lowest leakage. Sufficient information is available to predict silica leakage and exchange capacity for most conditions encountered in practice. Countercurrent operations allow Type II exchangers to reduce silica to the same degree as found with Type I resins, and still retain their high capacity efficiency when regenerated with caustic.

5

Specific Water Quality and Treatment Requirements

High quality water is used in a variety of applications at industrial facilities, for:

- feedwater for heating and power plants, including steam and water supply to ships

- cooling systems for air compressors, engines, heat exchangers, and for power and refrigeration plants

HEATING AND POWER PLANTS

System Components

Before considering boiler water treatment, it is helpful to focus briefly on the important functional components of a typical steam generation system. Figure 5-1 illustrates a typical medium pressure system (600 psig and less).

Treatment Equipment

Pretreatment varies with raw water characteristics, system operating pressure, required steam quality, average steam load, amount of

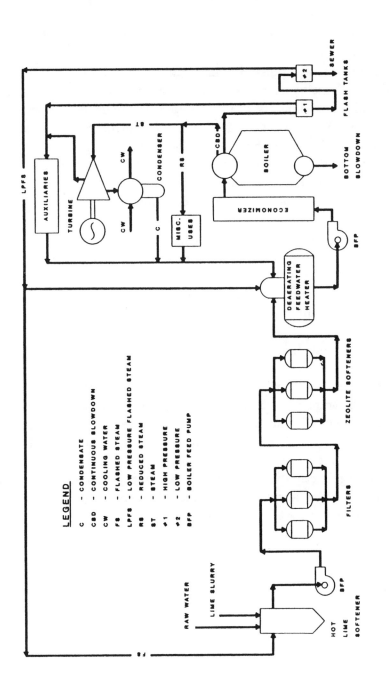

Figure 5-1. Steam Generating System for Power Generation and Heating

returned condensate, type of boiler, fuel costs, degree of heat recovery, and overall costs.

Pretreatment for low pressure boilers (under 100 psig) should be evaluated on a case-by-case basis, considering the above variables. Sodium zeolite softening is generally used for medium pressure boilers (100 to 600 psig). Occasionally either hot or cold process softening is used. For high pressure boilers (over 600 psig), demineralization is most often employed. Condensate polishing for iron and copper removal frequently is used in high pressure systems and in systems where no return-line corrosion control is either possible or desirable.

The primary objective of most pretreatment systems is reduction of boiler and superheater deposits. Reduced blowdown and return-line corrosion also are significant and common objectives. Contaminated condensate is often pretreated to enable recovery of waste heat and removal of impurities prior to return to the feedwater.

Deaerating Feedwater Heaters

Deaerating heaters, to remove dissolved gases, increase feedwater temperature by direct contact with low pressure exhaust steam. Most modern deaerating feedwater heaters include storage sections to hold heated deaerated boiler feedwater. Tanks are usually sized to hold sufficient water for ten minutes of plant operation at maximum load. The storage section is also a suitable place to introduce chemicals, such as oxygen scavengers, that require adequate mixing and retention time before water reaches the boilers.

Economizers

Hot flue gases discharged to the stack constitute the greatest single loss of heat in a fired boiler. An "economizer" is a heat exchanger placed in the gas passage between the boiler and stack, designed to recover waste heat from combustion products. Generally, a one percent increase in efficiency is realized for every 10°F to 11°F increase in feedwater temperature. Most modern economizers are constructed of steel, although cast iron is used when corrosion due to condensation of sulfuric or nitric acid in the exhaust is a consideration.

The waterside of an economizer is an extension of the boiler feedwater line. The increase in temperature increases the likelihood of both temperature deposition and corrosion. Accordingly, in systems with economizers, feedwater treatment must be both more carefully administered and be more extensive.

Where economizers are used, mechanical and chemical deaeration are essential. Deaerator performance should be checked regularly. Dissolved oxygen test samples should be taken (with the chemical oxygen scavenger omitted during the tests). Because, under certain inlet temperature and pressure conditions, air can be drawn into the feedwater pump through worn packing, samples should be taken downstream from the feedwater pumps.

Blowdown Systems

Blowdown is the purge from the system of a small portion of concentrated boiler water to limit concentration of dissolved and suspended solids.

Continuous surface blowdown offers the most economical and consistent control of total dissolved solids (TDS) or of any specific dissolved solid. Automatic control of continuous blowdown can be accomplished by using a conductivity monitor to actuate a blowdown control valve. Such devices increase economy since they provide control to a set point, avoiding the waste of water, fuel, and treatment chemicals that accompanies unnecessarily high blowdown rates.

Settled sludges must be removed by occasional manual bottom blowdown. In systems using hard water for boiler makeup, where suspended solids are the determining factor for cycles of concentration, manual blowdown may be used alone. Frequent manual blowdowns of short duration are more effective in sludge removal than occasional blowdowns of longer duration.

The blowdown from any boiler, whether intermittent manual bottom drum, purge, or continuous surface blowdown, is hot pressurized fluid. Special provisions are required for safe and efficient blowdown discharge. Sudden reduction of pressure causes flashing of some portion of the fluid to lower pressure steam. For safety, this phase change should occur in an enclosed vessel, or flash tank.

For low pressure heating boilers, flashed steam is vented safely to atmosphere, while cooled blowdown is discharged. But for most boilers operating at 100 psig or more, where flashing can occur at 5 psig or greater, the flash steam is recovered for use in the system. Such steam is typically used to preheat boiler feedwater in the deaerator.

Boilers

Modern subcritical boilers are either firetube or watertube. Operating pressure, capacity, and boiler efficiency vary over a considerable range for each type, although firetube designs are generally limited to less than 10,000 pounds per hour and to less than 250 psig steam pressure. Watertube boilers range in capacity up to several million pounds per hour with temperatures up to 1050°F at supercritical pressures. In general, watertube boilers require higher purity water than firetube units.

Almost all firetube boilers manufactured today are shop fabricated, packaged units, of design similar to the Scotch-type shown in Figure 5-2.

Firetube boilers are characterized by the heat exchange tubes being contained inside the drum. Design advances have led to very high heat release and heat absorption rates. Therefore, firetube boilers are highly susceptible to corrosion and to deposition on internal surfaces. Since the water is on the outside of the boiler tubes, the tube bundle must be chemically cleaned when necessary.

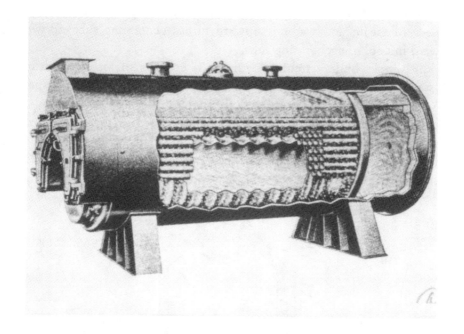

Figure 5-2. Scotch-Type Boiler

Source: Nebraska Boiler Co., Lincoln, NB

In watertube boilers, water is converted to steam inside the tubes, while hot gases pass on the outside. The tubes are interconnected to common water channels, or drums, and to steam outlets. Tube banks are generally constructed with a series of baffles. These lead combustion gases across heating surfaces to obtain maximum heat absorption. Watertube boilers are available in a wide range of designs that vary in operating pressure, capacity, quality of steam produced, type of fuel, and installation and start-up costs.

While most watertube boilers operate by natural circulation, some use pump induced forced circulation. Natural circulation is created by the difference in density between steam and water. In tubes generating steam, the mixture of steam and water has a lower density than the water in nongenerating tubes. Thus, in a boiler circuit, the flow of steam and water is upward in hot generating tubes, downward in cool nongenerating tubes.

A simple boiler circuit (Figure 5-3) consists of an upper drum (steam drum) and a lower drum (mud drum) connected by tubes. Steam generating tubes located in the hottest area of the boiler, where the combustion gases first pass, are called risers. Risers carry the steam-water mixture upward to the upper drum where steam is released. Water then flows from the upper drum through the cooler tubes, called downcomers, to the lower drum.

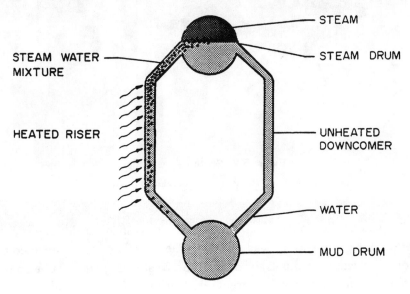

Figure 5-3. Simple Watertube Circuit

In a controlled, or forced circulation boiler, a circulating pump draws suction from a few large downcomer tubes and discharges into headers supplying the generating tubes. This design allows smaller diameter generating tubes and more intricate tube circuits.

In natural, or forced draft units, proper circulation is necessary for operation of a watertube boiler. As the steam-water mixture flows through the boiler tubes it cools tube metal, which is receiving heat from combustion gases. If circulation is inadequate, the boiler tubes will overheat and eventually fail.

The three basic designs of packaged, shop-fabricated watertube boilers manufactured are designated "D," "O," and "A" type. Types differ in arrangement of steam and mud drums, and in the path of combustion gases. A "D" type boiler is shown in Figure 5-4.

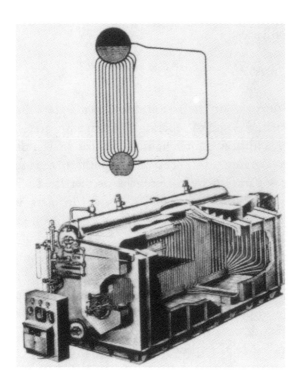

Figure 5-4. "D"-Type Boiler

Source: Erie City Boiler Co., Erie City, PA

The physical size and capacity of packaged watertube boilers is limited primarily by shipping restrictions from point of manufacture to point of delivery. Thus, field-erected watertube boilers usually are employed when steam capacity requirements exceed 200,000 pounds per hour.

Field-erected boilers usually consist of two drums connected by risers and downcomers positioned above and to the side of a relatively large radiant furnace, lined with waterwall tubes. Since the furnace releases considerable heat, the waterwall tubes are responsible for a significant portion of the steam generated. Field-erected units usually include auxiliary equipment such as economizers, air heaters, and superheaters. Packaged boilers also may be equipped with these auxiliaries.

Steam Drum Internals

The main purpose of the steam drum in a watertube boiler is to provide sufficient volume to separate steam from water. This is usually assisted by addition of mechanical devices in the drum, causing steam to travel greater distances on its path to the outlet header, and effecting separation by differences in inertia. Such devices reduce the mechanical entrainment of water droplets in the steam (mechanical carryover) but do not affect vaporous materials in the steam (volatile carryover).

Requirements for steam purity vary widely. As a general rule, highest purity is required when steam is used to drive turbines. This usually coincides with requirements for high-pressure boiler operation.

In smaller, low pressure boilers (firetube and watertube), elaborate steam-separating devices generally are not used. Baffles, screens, and dry-pipes installed in the space above water level, in the steam drum, are used to reduce mechanical entrainment.

Modern, high-pressure units require high purity steam to prevent deposit formation in superheaters and on turbine blades. The problem of designing efficient drum internals increases with pressure.

This is caused primarily by changes in the physical properties of water and of steam when temperature is increased.

While separation equipment has a marked effect on carryover, other factors are also significant. Carefully selected and controlled internal water treatment, plus pretreatment of feedwater, reduce carryover tendency of boiler water.

Superheaters

Saturated steam, leaving the drums of large boilers, commonly is directed through superheater sections before leaving the boiler. A superheater heats steam above saturation temperature. Superheated steam contains more energy than saturated steam, and, therefore, provides added driving potential for turbines with very little increase in fuel consumption.

Superheater tubes have steam on one side and hot combustion gases on the other. It is important to maintain steam flow in each superheater tube to avoid overheating. Operation at temperatures above design for the tube metal employed, even though not sufficiently high to result in tube failure, can result in excessive iron oxide formation inside the tube. This oxide tends to spall during temperature changes associated with startup and shutdown and may cause severe abrasion of a steam turbine nozzle block and first-stage blades. It is important that tube surfaces should be clean both internally and externally, and that carryover should be minimized. Otherwise, internal deposits of boiler water solids can result in overheating and possible failures.

Many superheater designs in modern boilers are the pendant, nondrainable type. In such systems, whenever the boiler is shut down, lower bends of each pass retain condensate produced from residual steam in the superheater. On startup, this condensate must be carefully re-evaporated and the steam released through a vent. During layup periods, or boiler cleanings, superheaters must be filled with high purity water containing a neutralizer and an oxygen scavenger.

Alternatively, the superheater can be purged with an inert gas such as nitrogen. Under no circumstances should a nondrainable superheater be filled with water containing nonvolatile dissolved solids, oxygen or carbon dioxide. The solids will deposit on superheater tube surfaces and will be very difficult to remove by flushing or chemical cleaning. The gases will corrode the metal. Where reheaters are employed, the same precautions apply.

Turbines

Turbines are rotary devices, driven by steam. The largest turbines are used to drive generators that produce electric power. In most units, the maximum energy in the steam is used to produce power before exhaust steam is condensed and returned as feedwater. In some steam-electric utility plants and in many industrial power plants, steam energy may be only partially used to generate electricity. The remaining energy, as lower pressure steam, is used in a steam distribution system for either heating or other purposes.

The economy and performance of a turbine depend on design and construction of turbine blades. Turbines are built with very close tolerances, and stationary and moving parts have little clearance. Any vibration has a deleterious effect. Of special concern is erosion of turbine blades by condensation. Deposits and erosion distort turbine blades and nozzle shape, producing rough surfaces and increasing steam flow resistance. Heavy blade deposits can cause turbine unbalance, producing intolerable vibration.

Generally, turbine deposits are caused either by carryover or corrosion. Nonselective boiler water carryover can be caused by a high level of total solids in boiler water, high boiler water alkalinity, poor feedwater quality (periodic contamination of some nature), and mechanical difficulties (steam separator failure, rapid load variation, high water level). Generally, deposits that form on turbine blades as a result of carryover can be rinsed from the blades with condensate formed during low load or unloaded operation.

Selective carryover is a more serious problem. In this case, one constituent is vaporized in the steam. In high pressure units, silica is the most common constituent carried over by vaporization. Silica volatility, at any pressure, is dependent on silica concentration and pH of the boiler water. Pure silica deposits are hard and glassy and cannot be removed by condensate. They are usually removed by sand or grit blasting. When sufficient nonselective carryover accompanies the volatile silica, soluble sodium silicate deposits may form.

Corrosion caused either by insufficient feedwater treatment, when the system is out of service, or by improper boiler and turbine layup procedure, can be a serious problem in turbines. Either partial or complete reblading is a frequent consequence of a failure to observe good out-of-service procedures.

Condensers

Exhaust steam leaving a turbine must either be used in the plant or be condensed before it re-enters the boiler. In industry there are two principal types of condensers. Each creates a vacuum as steam is condensed. The vacuum creates reduced back pressure on the turbine and greatly increases unit efficiency.

A surface condenser is most commonly used. It consists of a closed vessel filled with tubes. Cooling water flows through the tubes, while steam flows around the tubes. The coolant flowing through the tubes can either be used on a once-through basis or recirculated through a cooling tower. Condensate flows out at the base of the condenser to a "hot well" where it is ready to re-enter the water-steam cycle. Noncondensible gases and air that have entered through inleakage are removed by ejectors or vacuum pumps from a special air-removal section of the condenser. Air and oxygen removal not only reduces backpressure on the turbine, but also minimizes iron and copper corrosion in the preboiler system. Ammonia removed by steam jet air ejectors will redissolve in jet condenser condensate and will recycle if this condensate is returned to the main turbine condenser. Such recycling can build ammonia concentrations that, if oxygen is present, becomes corrosive to either copper or brass condenser tubes in the air removal section. Injection of catalyzed hydrazine can be beneficial when such problems occur. In such cases, jet condenser condensate should be handled separately. In addition to corrosion problems, condensers are subject to erosion and vibration on the steam side. Condensate droplets in exhaust steam, at sufficiently high velocity, can erode portions of the top two rows of tubing to failure. Shields or grids of more resistant alloy are placed either on or ahead of the first row of tubing to prevent failure by fretting and fatigue. Factors responsible are inlet steam velocity and excessive distance between tube support plates for the material, diameter, and wall thickness of the tubing used. Elimination of vibration in condensers often requires reducing turbine capacity.

Heating and Process Equipment

Central stations and closed heating systems are two common configurations that use steam generated in closed, noncontaminating equipment. All steam is condensed and returned to the boiler as feedwater. In such systems, there is very little loss of water. Consequently, the need for treated makeup water is minimized.

A great number of plants use steam for process equipment heating where condensate is not recoverable. This is often caused by absorption of the steam by the product or by its contamination during contact with the product. In other cases, condensate is discarded because collection and return is not economical. Such systems require large amounts of makeup water.

WATER TREATMENT OBJECTIVES

Boiler feedwater, regardless of the type and extent of external treatment, may contain contaminants that cause deposits, corrosion, and carryover.

Deposits directly reduce heat transfer, causing higher fuel consumption, high metal temperatures, and, eventually, failures. Deposits, although most serious in the boiler, may also cause problems in preboiler or after-boiler systems.

Corrosion not only results in failure at the point of attack, but also produces metal oxide contamination that may cause deposits elsewhere. Problems of deposit formation and corrosion are so closely related that both must be effectively corrected or prevented to achieve satisfactory treatment.

Boiler steam carryover results in superheater, turbine, and condensate system deposits, corrosion, and/or erosion problems. Very serious losses in efficiency, especially when superheated steam is used for power generation, result from loss of superheat, turbine blade deposits, erosion, or corrosion. Even extremely low levels of carryover may cause failures and complete system outages.

Therefore, the three main objectives of boiler water treatment are to:

- Prevent formation of deposits
- Reduce corrosion of metals
- Prevent boiler water carryover

All parts of the steam-water system are interdependent. In deciding on treatment to be used, the entire system must be considered.

Tables 5-1A through 5-1D contain recommended limits for water quality as a function of boiler pressure. These limits, set by an ASME subcommittee, differ from the limits specified by the American Boiler Manufacturers Association (ABMA) in Table 5-2. The ABMA specifies limits for boiler water composition to assure good quality steam. There are other considerations, such as corrosion and deposit formation, in establishing boiler water composition. With today's designs, areas of watertube boilers can reach heat-flux rates far greater than those of earlier boilers. This has resulted in the need for new guidelines to replace the outdated ABMA limits.

Table 5-1A. ASME Industrial Boiler Subcommittee Suggested Water Quality Limits

Boiler Type: Industrial watertube, high duty, primary fuel fired, drum type
Makeup Water Percentage: Up to 100% of feedwater
Conditions: Include superheater, turbine drives, or process restriction on steam purity
Saturated Steam Purity Target[9]

Drum Operating Pressure (psig)[1]	0-300	901-450	451-600	601-750	751-900	901-1000	1001-1500	1510-2000
Feedwater[7]								
Dissolved oxygen (mg/L O_2) measured before oxygen scavenger addition[8]	<0.04	<0.04	<0.007	<0.007	<0.007	<0.007	<0.00	<0.007
Total iron (mg/L Fe)	<0.100	<0.050	<0.030	<0.025	<0.020	<0.020	<0.010	<0.010
Total copper (mg/L Cu)	<0.050	<0.025	<0.020	<0.020	<0.100	<0.050	<0.10	<0.010
Total hardness (mg/L $CaCO_3$)	<0.300	<0.300	<0.200	<0.200	<0.100	<0.050	- - - Not Detectable - - -	
pH range @ 25°C	7.5-10.0	7.5-10.0	7.5-10.0	7.5-10.0	7.5-10.0	8.5-9.5	9.0-9.6	9.0-9.6
Chemicals for preboiler system protection							Use only volatile alkaline materials	
Nonvolatile TOC (mg/L C)[6]	<1	<1	<0.5	<0.5	<0.5	<0.5	. . . As low as possible, <0.2 . . .	
Oily matter (mg/L)	<1	<1	<0.5	<0.5	<0.5	<0.5	. . . As low as possible, <0.2 . . .	
Boiler Water								
Silica (mg/L SiO_2)	<150	<90	<40	<30	<20	<8	<2	<1
Total alkalinity (mg/L $CaCO_3$)	<350[3]	<300[3]	<250[3]	<200[3]	<150[3]	<100[3]	- - - Not Specified - - -	
Free hydroxide alkalinity (mg/L $CaCO_3$)	- - - - - - - - - - - - Not Specified - - - - - - - - - - - -						- - Not Detectable[4] - -	
Specific conductance (uS/cm) (umho/cm) @ 25°C without netralization	<3500[5]	<3000[5]	<2500[5]	<2000[5]	<1500[5]	<1000[5]	<150	<100

Notes for Table 5-1A. ASME Industrial Boiler Subcommittee Suggested Water Quality Limits

1. With local heat fluxes >150,000 Btu/hr/ft^2, use values for the next higher pressure range.

2. Minimum level of OH$^-$ alkalinity in boilers below 900 psig must be individually specified with regard to silica solubility and other components of internal treatment.

3. Maximum total alkalinity consistent with acceptable steam purity. If necessary, should override conductance as blowdown control parameter. If makeup is demineralized water at 500 psig boiler water alkalinity and conductance should be that in table for 1,001 to 1,500 psig range.

4. Not detectable in these cases refers to free sodium or potassium hydroxide alkalinity. Some small variable amount of total alkalinity will be present and measurable with the assumed congruent or coordinated phosphate—pH control or volatile treatment employed at these high pressure ranges.

5. Maximum values often not achievable without exceeding suggested maximum total alkalinity values, especially in boilers below 900 psig with >20% makeup of water whose total alkalinity is >20% of TDS naturally or after pretreatment by lime-soda, or sodium cycle ion exchange softening. Actual permissible conductance values to achieve any desired steam purity must be established for each case by careful steam purity measurements. Relationship between conductance and steam purity is affected by too many variables to allow its reduction to a simple list of tabulated values.

6. Nonvolatile TOC is that organic carbon not found intentionally added as part of the water treatment regime.

7. Boilers below 900 psig with large furnaces, large steam release space and internal chelant, polymer, and/or antifoam treatment can sometimes tolerate higher levels of feedwater impurities than those in the table and still achieve adequate deposition control and steam purity. Removal of these impurities by external pretreatment is always a more positive solution. Alternatives must be evaluated as to practicality and economics in each individual case.

8. Values in table assume existence of a deaerator.

9. No value given because steam purity achievable depends upon many variables, including boiler water total alkalinity and specific conductance as well as design of boiler, steam drum internals, and operating conditions (Note 5). Since boilers in this category require a relatively high degree of steam purity, other operating parameters must be set as low as necessary to achieve this high purity for protection of the superheaters and turbines and/or to avoid process contamination.

Table 5-1B. ASME Industrial Boiler Subcommittee Suggested Water Quality Limits

Boiler Type: Industrial watertube, high duty, primary fuel fired, drum type
Makeup Water Percentage: Up to 100% of feedwater
Conditions: No superheater, turbine drives, or process restriction in steam purity
Saturated Steam Purity Target[7]: 1.0 mg/L (ppm) TDS maximum

Drum Operating Pressure (psig)	0-300	301-600
Feedwater[3]		
Dissolved oxygen (mg/L O_2) measured before chemical oxygen scavanger addition[1]	<0.04	<0.04
Dissolved oxygen (mg/L O_2) measured after chemical oxygen scavanger addition[2]	<0.007	<0.007
Total iron (mg/L Fe)	<0.10	<0.050
Total copper (mg/L Cu)	<0.05	<0.025
Total hardness (mg/L as $CaCO_3$)	<0.5	<0.3
pH range @ 25°C	7.0-10.5	7.0-10-5
Nonvolatile TOC (mg/L C)[6]	<1	<1
Oily matter (mg/L)	<1	<1
Boiler Water		
Silica (mg/L SiO_2)	<150	<90
Total alkalinity (mg/L as $CaCO_3$)	<1,000	<850[5]
Free hydroxide alkalinity (mg/L as $CaCO_3$)[4]	------ Not Specified ------	
Specific conductance (uS/cm) (umho/cm) at 25°C without neutralization	<8,000[5]	<6,500[5]

NOTES for Table 5-1B. ASME Industrial Boiler Subcommittee Suggested Water Quality Limits

1. Value in table assumes existence of a deaerator.

2. Chemical deaeration must be provided in all cases but especially if mechanical deaeration is nonexistent or inefficient.

3. Boilers with relatively large furnaces, large steam release space and internal chelant, polymer, and/or antifoam treatment can often tolerate higher levels of feedwater impurities than those in the table and still achieve adequate deposition control and steam purity. Removal of these impurities by external pretreatment is always a more positive solution. Alternatives must be evaluated as to practicality and economics in each individual case. The use of some dispersant and antifoam internal treatment is typical in this type of boiler operation so it can tolerate higher feedwater hardness than the boilers in Table 5-1A.

4. Minimum level of OH$^-$ alkalinity must be individually specified with regard to silica solubility and other components of internal treatment.

5. Alkalinity and conductance values consistent with steam purity target. Practical limits above or below tabulated values can be established for each case by careful steam purity measurements.

6. Nonvolatile TOC is that organic carbon not intentionally added as part of the water treatment regime.

7. Target value represents steam purity which should be achievable if other tabulated water quality values are maintained. The target is not intended to be nor should it be construed to represent a boiler performance guarantee.

Table 5-1C. ASME Industrial Boiler Subcommittee Suggested Water Quality Limits

Boiler Type: Industrial firetube, high duty, primary fuel fired
Makeup Water Percentage: Up to 100% of feedwater
Conditions: No superheater, turbine drives, or process restriction on steam purity
Saturated Steam Purity Target[7]: 1.0 mg/L (ppm) TDS maximum

Operating Pressure (psig)	0-300
Feedwater[3]	
Dissolved oxygen (mg/L O_2) measured before chemical oxygen scavenger addition[1]	<0.04
Dissolved oxygen (mg/L O_2) measured after chemical oxygen scavenger addition[2]	<0.007
Total iron (mg/L Fe)	<0.10
Total copper (mg/L Cu)	<0.05
Total hardness (mg/L as $CaCO_3$)	<0.5
pH range @ 25°C	7.0-10.5
Nonvolatile TOC (mg/L C)[6]	<1
Oily matter (mg/L)	<1
Boiler Water	
Silica (mg/L SiO_2)	<150
Total alkalinity (mg/L as $CaCO_3$)	<700
Free hydroxide alkalinity (mg/L as $CaCO_3$)[4]	Not Specified
Specific conductance (uS/cm) (umho/cm) at 25°C without neutralization	<7,000[5]

NOTES for Table 5-1C. ASME Industrial Boiler Subcommittee Suggested Water Quality Limits

1. Value in table assumes existence of a deaerator.

2. Chemical dearation must be provided in all cases but especially if mechanical deaeration is nonexistent or inefficient.

3. Firetube boilers of conservative design, with internal chelant, polymer and/or antifoam treatment can often tolerate higher levels of feedwater impurities than those in the table (<0.5 mg/L Fe, <0.2 mg/L Cu, <10 mg/L total hardness) and still achieve adequate deposition control and steam purity. Removal of these impurities by external pretreatment is always a more positive solution. Alternatives must be evaluated as to practicality and economics in each individual case.

4. Minimum level of OH⁻ alkalinity must be individually specified with regard to silica solubility and other components of internal treatment.

5. Alkalinity and conductance guidelines consistent with steam purity target. Practical limits above or below tabulated values can be established for each case by careful steam purity measurements.

6. Nonvolatile TOC is that organic carbon not intentionally added as part of the water treatment regime.

7. Target value represents steam purity which should be achievable if other tabulated water quality values are maintained. The target is not intended to be nor should it be construed to represent a boiler performance guarantee.

Table 5-1D. ASME Industrial Boiler Subcommittee Suggested Water Quality Limits

Boiler Type: Industrial, coil type, watertube, primary fuel fired rapid steam generators
Makeup Water Percentage: Up to 100% of feedwater
Total Evaporation: Up to 95% of feedwater
Steam to Water Ratio (volume to volume): Up to 4,000 to 1
Saturated Steam Purity Targets: See Table

Operating Pressure (psig)	0-300	301-450	451-600	601-900	901 and up
Steam Purity Targets[1]					
Specific conductance @ 25°C (uS/cm) (umho/cm)	≤50	≤24	≤20	≤0.5	≤0.2
Dissolved solids (mg/L)	≤25	≤12	≤10	≤0.25	≤0.1
Silica (mg/L SiO_2)	— —	— —	— —	<0.003	<0.002
Feedwater[2]					
Dissolved oxygen (mg/L O_2) measured before chemical oxygen scavenger addition[8]	<0.2	<0.2	<0.2	<0.007	<0.007
Dissolved oxygen (mg/L O_2) measured after chemical oxygen scavenger	<0.007	<0.007	<0.007	<0.003	<0.003
Total iron (mg/L Fe)	<1.0	<0.3	<0.1	<0.05	<0.02
Total copper (mg/L Cu)	<0.1	<0.05	<0.03	≤0.02	≤0.02
Total hardness (mg/L $CaCO_3$)	<1.0	<0.7	<0.5	- - - Not Detectable - - -	
pH range @ 25°C	9.0-9.5	9.0-9.5	8.8-9.2	8.8-9.2	8.8-9.2
Chemicals for preboiler system protection			Use only volatile alkaline materials		
Boiler Water					
Silica (mg/L SiO_2)	≤150	≤100	≤60	≤30	≤10[7]
Total alkalinity (mg/L $CaCO_3$)	<800[3]	<600[3]	<500	<200	<100[7]
Hydroxide alkalinity (mg/L $CaCO_3$)[8]	>300	>200	>120	>60	≤50[7]
Specific conductance (uS/cm) (umho/cm) @ 25°C without neutralization	<8,000	<6,000	<5,000	<4,000	<500[7]

NOTES for Table 5-1D. ASME Industrial Boiler Subcommittee Suggested Water Quality Limits

1. Tabulated values based on assumption of no superheaters or turbine drives. If steam used for superheat or turbine drives, use values for 901 psig and up. If unit operation approached superheat conditions within coil, use values for 601 to 900 psig range to avoid silica deposition on near-dry surfaces. The target is not intended to be nor should it be construed to represent a boiler performance guarantee.

2. Feedwater defined as makeup water plus condensate, other than separator returns.

3. Values in table assume existence of a deaerator.

4. Chemical deaeration with catalyzed oxygen scavenger is necessary in all cases because feedwater temperature limits imposed by manufacturers of coil type steam generators preclude efficient mechanical deaeration. Feed of chemical oxygen scavenger must be sufficient to maintain a detectable residual in the boiler water. For those units which include steam separator-water storage drums and recirculate substantial amount of boiler water, oxygen scavenger residuals should be maintained in higher ranges typical of those employed for drum type boilers.

5. Demineralization of makeup water recommended.

6. Boiler water analyses determined on separator discharges and/or on storage drum sample.

7. Suggested values vary with operating pressure decreasing proportionally as pressure increases up to 2,500 psig.

8. Hydroxide alkalinity in mg/L $CaCO_3$ must be maintained at a minimum of 1.7 times silica in mg/L SiO_2 to keep silica soluble and avoid complex silicate deposits. Most coil type steam generators do not employ scale control internal treatment chemicals to assist in prevention of such deposits.

BOILER SYSTEM PROBLEMS

Carryover of Boiler Water

Maintenance of steam purity is of extreme importance for plant operation. The presence of small quantities of inorganic salts in water carried over from the boiler to the steam can cause superheater failures, loss of turbine efficiency, and related problems. This can also increase potential for corrosion or erosion-corrosion in steam condensate systems. In addition, carryover may cause contamination in direct steam use processes.

In steam generators equipped with superheaters, carryover may be serious because materials carried over may form deposits in the superheater, resulting in overheating and subsequent tube failures. Solids concentrations in boiler water greater than 1.0 ppm will cause superheater deposits that may form at concentrations in the steam as low as 0.1 ppm. Steam purity is essential when turbines follow superheaters.

Carryover of dissolved material, from boiler water to steam, is caused by entrainment of small droplets of boiler water in steam leaving the drum and by volatilization of dissolved salts that are soluble in steam.

Mechanical entrainment is a function of both design and method of operation of steam separators in the boiler. Conditions that produce foaming increase the presence of droplets in the steam separating equipment. The entrainment process may be divided into two categories—priming and foaming. Priming usually results from a sudden reduction in boiler pressure caused by a rapid increase in the boiler steam load. This causes steam bubbles to form throughout the mass of water in the boiler. Increased water volume raises the level in the drum, flooding the separators or dry pipe. Priming may also result from excessively high water level following a rapid load reduction caused by control failure. Foaming is a buildup of bubbles on the water surface in the steam drum. This reduces the steam-release space and, by various mechanisms, causes mechanical entrainment. Mechanical entrainment is usually controlled through proper design and efficient operation, supplemented by the use of antifoam agents in situations when foaming becomes a problem.

Design Considerations

The steam drum of any boiler must be designed with sufficient volume to allow separation of water from steam before the steam leaves the boiler.

Boiler drums may contain baffles, screens, mesh demisters, chevron separators, and/or centrifugal separators to improve separation of water droplets from steam. Each element must be kept tight and clean; a quarter-inch gap between the sections of cover baffles over the generating tubes can cause water to bypass separators and thereby negate their operation. In similar fashion, the presence of deposits on screens or mesh demisters can prevent the devices from functioning properly. It is essential for all drum components to be inspected regularly and maintained in proper working order.

Mechanical entrainment is rarely uniform along the length of a drum. Any sampling technique used for monitoring steam purity must take this into account and provide for collection from a variety of points in the drum. Steam samples taken from the discharge pipe of the boiler drum should be taken by the use of an isokinetic sample probe, to obtain a representative sample.

Operational Effects

Since size of a steam drum is fixed, operating water level determines the vapor-liquid separation space. When this level is excessively high, separation area and volume are both minimized and continuous entrainment can result. Lowering operating water level can correct this situation. Operation below mid-drum level should be evaluated carefully because while steam separation volume is increased, disengagement area is reduced. In addition, angle and point of entry of the steam-water mix from the generating tubes is important.

Mechanical carryover from boilers is frequently caused by operation at steam generation loads exceeding design rating. This may occur either continuously or intermittently as a result of sudden process steam demands. Load surges cause a sudden lowering of pressure in the drum, with resultant violent boiling and discharge of boiler water into the steam release space.

Uneven fuel firing may cause excessive localized steam generation and result in carryover, even when other operational aspects are normal. Where carryover is the result of foaming, various steps including use of antifoam agents can correct the problem.

Chemical Considerations

Separation of water droplets from steam is a function of the surface tension of the boiler water. Surface tension affects the size of steam bubbles formed on the heat transfer surface and the ease with which they coalesce and collapse. Lower surface tension results in formation of small bubbles and tends to stabilize them in the steam drum, causing an excessive quantity of small droplets of boiler water to enter the final stages of the steam separating equipment.

The components in boiler water that can increase potential for mechanical entrainment include alkalinity, suspended solids, dissolved solids, and organic surfactants such as saponified oils and synthetic detergents. A program to control the chemistry of boiler water must regulate these components. Since entrainment increases with operating pressure, TDS must decrease with increasing pressure to maintain a given steam purity. This is illustrated by boiler water quality standards established by the American Boiler Manufacturers Association (ABMA), Table 5-2. These figures, of an older standard, are not restrictive enough for high-tech boilers.

The total dissolved solids level is usually determined by conductivity measurements on a neutralized sample of boiler blowdown water and is often read directly as micromhos of specific conductance. This level is controlled in boiler water primarily by varying the amount of surface blowdown. Careful selection of chemicals can minimize dissolved solids in boiler water. It is often economically desirable to reduce boiler water solids by pretreating makeup water, and by returning a greater percentage of condensate.

Table 5-2. American Boiler Manufacturers Association Water Quality Standards

Boiler Pressure (psig)	PPM Total Solids	Total Alkalinity PPM as CaCO₃	PPM Suspended Solids	Silica* (ppm as SiO₂)
0 to 300	3500	700	300	125
301 to 450	3000	600	250	90
451 to 600	2500	500	150	50
601 to 750	2000	400	100	35
751 to 900	1500	300	60	20
901 to 1000	1250	250	40	8
1001 to 1500	1000	200	20	2.5
1501 to 2000	750	150	10	1.0
Over 2000	500	100	5	0.5

*Silica limits based on limiting silica to steam to 0.02 to 0.03 ppm

The total alkalinity of boiler water is a function of makeup water alkalinity and of the alkalinity of treatment chemicals added for internal deposit or corrosion control. Makeup water alkalinity is usually in the form of carbonates and bicarbonates, which decompose at boiler temperature to release carbon dioxide and form hydroxide alkalinity. The alkaline treatment chemicals are caustic soda and soda ash.

Alkalinity can be controlled in boiler water by increasing blowdown, by external dealkalization of makeup water, usually by lime treatment, ion exchange on the hydrogen cycle or decarbonization by acid addition, or by elimination of alkaline chemical feed. The most effective and economical method is external treatment of makeup water, because such treatment often produces an additional benefit by reducing total dissolved solids.

Suspended solids can be present in boiler water as a result of precipitation of boiler sludges, suspended matter carried in with feedwater, or corrosion products. Possible corrective measures are control of formation and concentration of suspended matter, removal of suspended solids from feedwater, internal treatment with a dispersant/antifoam agent and/or minimizing corrosion products by either condensate polishing or improved corrosion control.

Use of Chemical Antifoams

Frequently, chemicals are added to boiler water to counteract effects of high TDS, alkalinity, suspended solids, oil or other organics and to allow operation without unacceptable chemical carryover. Such a program may be more economical than either increased blowdown or additional external treatment. Occasionally the use of a chemical antifoam agent is necessary to operate a boiler without gross entrainment. Chemical antifoams frequently reduce fuel consumption by permitting lower blowdown rates while producing high quality steam.

Polyglycols and polyamides are chemicals most often employed as antifoam agents. Of these, polyglycols have been the most successfully used and the most widely accepted. They are fed directly to the boiler or to the feedwater line in low dosages and are often blended with other internal treatment chemicals.

In most foaming situations, controlled use of an antifoam agent will successfully reduce carryover. There are instances where the chemical agent provides no improvement and may even increase carryover. The decision to use an antifoam agent should be made strictly on the basis of steam quality measurements.

Volatile Carryover

When steam is generated at high pressure many solids dissolved in boiler water volatilize and are carried from the boiler as dissolved matter in steam. The term for such contamination of steam is volatile carryover. These contaminants can deposit in turbines at any point where temperature conditions allow condensation.

Volatile carryover cannot be prevented by either mechanical or operational modifications, or by addition of chemical antifoam agents. The only remedy is to limit the concentration of volatile solids in the boiler water, or to vary pH, a factor that influences volatility.

Volatile carryover begins to reach significant levels in boilers operating at 600 psig and, thereafter, increases proportionately with pressure. The most important volatile solid is silica, although the volatilities of sodium hydroxide, sodium chloride, and sodium phosphate can also reach significant levels, especially at higher pressures and concentrations. Vaporization of dissolved materials such as ammonia, carbon dioxide, hydrogen sulfide, sulfur dioxide, and morpholine, which are normally gaseous or highly volatile, also occurs. With the exception of ammonia and morpholine, these materials are usually absent in high pressure boiler systems.

Silica can vaporize and deposit on turbine blades at boiler pressures greater than 600 psig. Therefore, its presence is cause for concern when steam is used for turbine drives. It is generally accepted that steam with 20 ppb, or less, silica will not result in silicon dioxide deposits on turbine blades. When sufficient sodium is present, however, complex sodium silicates may deposit at concentrations lower than 20 ppb.

Iron and copper contamination of steam can be reduced by lowering concentration of these materials in the feedwater. Catalyzed hydrazine has been highly effective in reducing pickup of iron and copper in lower temperature sections of central stations.

Deposition from Steam

Deposits of boiler water salts can occur in turbines, whether carryover is mechanical, volatile, or both.

Silica is one of the more volatile salts commonly encountered in boiler water. Its solubility in superheated steam decreases rapidly as steam temperature and pressure decrease. Deposits form as the silica is removed from solution. Deposits in superheaters and high temperature steam lines are caused by mechanical entrainment or contaminated desuperheating water.

Deposit Removal from Superheaters

Once carryover deposits have formed in a superheater, removal is exceedingly difficult and dangerous. In a drainable superheater, it is possible to fill the system with hot boiler water on shutdown, or to backfill with hot condensate from the outlet header. Soaking with either of these waters should dissolve deposits unless the accumulation is such that tubes are literally plugged. After the soak, the unit is drained and then flushed with high purity water until effluent conductivity indicates complete removal of dissolved salts.

In nondrainable superheaters, the procedure is more complicated, because it is nearly impossible to insure adequate flow through all lines. Backfill from outlet headers or manual flushing of individual tubes from the drum is possible, but the necessary volume of high pressure, high-purity water usually is not available. It is possible to flood the superheater with either hot, high-purity feedwater or condensate after taking the unit out of service and before internal pressure returns to ambient. High-purity water can then be used to flush the superheater once the steam has condensed and all loops are filled with water. Sufficient flushing must take place to assure that the superheater is free of all deposits.

Acid solvent cleaning, when deposits are water insoluble, is not recommended in nondrainable superheaters, because potential exists for greater damage in cleaning than would occur by leaving deposits in place.

Removal of Deposits from Turbines

Water soluble deposits can be removed from turbine blades by condensate-washing either during operation or on shutdown. This involves removing the load from the turbine but allowing the machine to spin on saturated steam. If the machine is full-condensing and can be removed from service intermittently, soluble deposits can be removed by periodic washing before they build to dangerous levels. Care must be taken to avoid damage to the turbine. Specific procedures must be followed closely. Caustic can be added to improve silica removal, but this procedure results in loss of large quantities of condensate. Specific conductance and/or silica are monitored in the condenser.

Some deposits, particularly silica, that form in turbines are not water soluble. In this case, the cleaning procedure is to open the casing, remove the rotor, and clean the blades mechanically.

Standby Protection Against Corrosion (Layup)

During nonoperational periods, protection against corrosion of boilers, auxiliary systems, and other related industrial equipment is essential. Downtime corrosive attack can cause loss of metal. Corrosion products are released from the preboiler section during service shutdowns. During subsequent periods of operation, iron and copper oxides formed by out-of-service corrosion can be transported to boiler heating surfaces. These oxides form deposits and may cause localized attack and tube metal overheating.

The history of operational corrosion in power station units and industrial boilers indicates that application of effective corrosion control measures during startup and shutdown periods, combined with continuous onstream controls, constitutes sound preventive maintenance. These measures and controls assist in protection of the original investment and extend the useful life of major plant equipment.

The key factors affecting downtime corrosion are water, oxygen, and pH. Elimination of either moisture or air will prevent appreciable corrosion. Dry layups accomplish this goal by eliminating water and by reducing the relative humidity to a safe level. Wet layups control corrosion by excluding oxygen and maintaining a high pH level. Layups using nitrogen gas to displace air from the boiler also control corrosion by excluding oxygen.

The decision to use wet, dry, or nitrogen gas standby protection depends on the duration of the downtime period and the degree of availability required of power plant equipment. Wet storage is employed regardless of duration if the boiler is to be kept available for immediate use. Dry storage is usually preferred for long shutdowns. Steam plant shutdowns of one month or less are considered short-term layup; those exceeding a month are regarded as long-term layup.

Short-Term, Wet Standby Protection

Wet layup techniques should be implemented for shutdowns lasting a month or less, and in situations requiring immediate availability. During wet storage, dual mechanisms used to protect the equipment are exclusion of oxygen and maintenance of high pH.

For boilers with softened feedwater and nondrainable superheaters, wet layup can only be accomplished by nitrogen blanketing both the superheater and the space above normal drum water level. For wet layup of high pressure boilers, the inhibitors used are hydrazine and ammonia or amines to adjust pH. Nitrogen blanketing should be employed if it is necessary to avoid flooding the superheater. Special layup procedures must be followed to protect idle units after acid cleaning. Ideally, chemical cleaning of a boiler with acidic solvent should be scheduled to make certain that boiler startup can proceed immediately after completion of cleaning. This includes neutralization, passivation, and inspection. Should a delay of more than one or two days occur between the end of cleaning and startup, the boiler should be filled with treated feedwater or condensate.

When filling a boiler for wet layup, it is desirable for corrosion inhibitors (proportioned to deaerated feedwater) to be fed into the economizer where they remain to afford standby protection.

The same wet standby treatment for a boiler is used for the superheater section, with one notable exception: caustic and sulfite cannot be used in nondrainable superheater layup. If wet layup of a nondrainable superheater is necessary, the recommended chemical inhibitors are hydrazine (or catalyzed hydrazine) and ammonia or neutralizing amine (morpholine, cyclohexylamine). This volatile treatment assures that no deposits are formed in the superheater when the boiler resumes operation.

Air ingress can cause exfoliation (a special type of oxidation in feedwater heaters during shutdown). Surface oxidation of hot tubes occurs on startup and shutdown.

Exfoliating corrosion can be controlled by methods other than use of alternative metals. Two such standby techniques are:

- Blanketing with inert gas or steam. This excludes oxygen, the source of the problem. Because air entering the system during shutdown is the only source of oxygen, the turbine, feed lines, and heater shell should be blanketed with nitrogen or steam. For most power plants, steam blanketing is more practical than nitrogen pressurization.

- Direct injection of octadecylamine, a filming amine corrosion inhibitor, into the feedwater heater shell.

When turbines are out of service, they must be kept dry using one or more of the following precautions:

- Tightening the block valve on the inlet steam line
- Installing a double block valve with a drain in between
- Bleeding either heated or dry instrument air into the turbine
- Following manufacturer's specifications for out-of-service storage

Long-Term Storage

For shutdowns either exceeding one month or those involving boilers subject to freezing conditions, dry storage techniques may be required to protect power plant equipment. Dry techniques fall into two general categories: open and closed.

Elimination of all moisture is the primary objective in dry standby, because as long as the boiler and other metal surfaces remain dry, no significant corrosion can occur. Air in contact with moisture is extremely corrosive and water leakage into the unit or sweating of surfaces must be avoided. The boiler should be drained completely, thoroughly cleaned on both fireside and waterside, and inspected. All internal surfaces should be dried. During dry storage, the boiler should be inspected for condensation. The success of dry layup is dependent on leak-free nonreturn valves, feedwater valves, and blowdown valves, when connected to an operating system.

Several different or combined treatments may be applied to an idle boiler stored dry and sealed. Although this approach can be effective if properly implemented, it presents some problems. Maintenance of air-tight seals on all boiler openings is difficult. Inspections during the shutdown period admit moist air. Desiccants, porous chemical drying agents, must be added to the secured boiler to absorb water vapor to control relative humidity. If the desiccant is caked or spent, failure to detect this condition during inspection may render the boiler subject to corrosion by humid air. Nonreturn valves must be leak-free. This type of dry storage is useful when the plant atmosphere contains corrosive fumes and/or abrasive dust, which could enter the open and dry boiler.

Another dry layup technique consists of completely sealing the drained boiler and maintaining a positive pressure of nitrogen. The nitrogen capping (blanketing) method is discussed under short-term, wet standby procedures.

Finally, extended storage of a boiler can also be accomplished by maintaining the boiler sealed and drained while under steam pressure.

Traps on the mud drum and lowest header remove condensate. Vents for noncondensibles are installed at dead-end points in the boiler. Dampers and all openings to the furnace, are closed tight. This system has the advantage of keeping the boiler hot, preventing condensation and resultant fireside corrosion.

DEPOSITS IN STEAM GENERATING SYSTEMS

Preboiler Section

The preboiler section includes all piping and equipment from the point of raw water pickup (river, wells, city water) to the point of entry into the boiler.

Preboiler section deposits can form in several locations. Deposits in deaerating heaters interfere with water distribution and deaeration, in feedwater lines restrict flow, in economizers and other feedwater heaters limit heat transfer while promoting corrosion, in feedwater regulators cause a loss of water level control, producing irregularities in circulation and causing carryover.

Preboiler section deposits may be caused by any of the following: oversaturation of calcium carbonate, reaction between hardness and treatment chemicals, contaminated condensate, presence of iron.

Deposits Caused by Contaminated Condensate

Contamination of steam and condensate generally originates from one or more of the following sources: leakage coils or jackets, backup of product in open coils, corrosion of metals in the afterboiler section, raw water injection into condensate to prevent flashing in vacuum return pumps or for desuperheating, condenser leakage, boiler carryover.

Steam can become contaminated with oily matter either when used in fuel oil heaters or when operating reciprocating pumps, compressors, and engines. Condensate from fuel oil heating is usually discarded. In many instances, baffle-type oil separators remove oil from the exhaust steam of equipment such as steam-driven reciprocating pumps. Separators must be checked periodically to ensure against malfunction. Process contamination occurs more frequently in intermittent operations than in continuous ones. The potential for contamination is greater when steam contacts products directly through open lines than in closed piping or jackets; leaks in either piping or jackets can result in contamination.

Contaminated condensate recovered and returned for use as feedwater may cause contaminants to react with other feedwater constituents. Precipitates may form deposits on metal surfaces. The contaminant may serve as a binder for other precipitates, making them more adherent. Condensate contamination caused by carryover of boiler water not only places sludge in the feedwater but also adds caustic and orthophosphate to the feedwater, which could result in additional sludge formation.

Frequent testing of condensate, either intermittent or continuous, is necessary to detect contamination. Steam and condensate conductivity alarm systems may be adequate for continuous monitoring. Such systems generally consist of a sample cooler, conductivity cell and accompanying holder, controller, and alarm. In some situations, a conductivity cell may be installed directly in a condensate line.

When impurities are detected, condensate should either be diverted or be discarded until the source of contamination has been located and eliminated. Other monitoring methods include sodium and hardness analyzers and measurements of pH, total organic carbon, and turbidity.

Deposition of Corrosion Products

Iron and copper corrosion products, in returned condensate, may cause preboiler section deposits either by direct adherence or by acting as a binder. Condensate should be monitored for metal ion content and corrosivity and for presence of condensate corrosion inhibitors. When inhibitors cannot be used, substitution of condensate filters and polishers to remove iron and copper oxides is appropriate.

Boiler Deposits

The most severe effects of deposits in steam generating systems occur in the boiler. The presence of deposits on heat transfer surfaces disrupts a boiler's basic function. Deposits on the waterside of a boiler result in overheating of tubes and subsequent metal softening, expansion, thinning, and, ultimately, tube failure. Boiler metal begins to lose strength rapidly as temperature increases above 900°F.

Deposit-caused tube failure usually occurs in sections of the boiler where the greatest heat transfer takes place. Riser (generating) tubes are, therefore, most susceptible. Boiler deposits result in tube failure when the insulating effect of a deposit elevates temperatures to the metal softening point. The type and thickness of a deposit determines the extent of reduced heat transfer.

Even in situations where deposits may not cause tube failure, their insulating effect results in energy waste.

Disruption of boiler circulation may occur when deposits loosened by thermal shock collect in a tube bend or when deposits build up on orifices in a forced circulation boiler. Carryover can result from deposit interference with steam separating devices, particularly where deposits build up on chevron or mesh demisters.

Tube failure, carryover, and other operating problems may make deposit removal necessary.

Control Programs

Boiler water treatment programs normally combine use of various antiscalants; the goal is to keep the boiler free of deposits. Selection of a treatment program depends on boiler design, operating pressure, makeup water constituents, type of external treatment, and method of operation. Deposit control schemes frequently include a precipitating program.

Boiler antiscalants can be divided into two categories: those that chemically react with feedwater impurities to change chemical structure, and those that alter the action of impurities. Commonly used reactants are carbonates, phosphates, and chelants. Antiscalants that alter behavior of the impurities are select "organics," polymers, and threshold-acting sequesterants.

In precipitating programs chemical treatment results in precipitation of boiler sludges instead of scale formation. Phosphate-based programs are precipitating programs. In phosphate programs, resulting sludges are calcium hydroxyapatite and serpentine. Both are relatively nonadherent and are removed from the boiler with bottom blowdown. Efficient bottom blowdown is essential because sludge may be swept from the bottom of the boiler (where it should settle) into hot sections of the boiler where it can bake on hot tube surfaces. The frequency of bottom blowdown is dictated by both the solids load on the boiler and the boiler design.

One type of solubilizing program employs polymers, metal dispersants, and organic sludge conditioners. These prevent scale formation by distorting crystalline growth of scale and dispersing small particles as colloids. Organics condition the sludge that is formed. This program relies quite heavily on use of specific metal dispersants to prevent binding of precipitates.

The choice of program is often dictated by economics. A chelant program, for example, is usually too expensive if hard water makeup is used.

Other Boiler Deposit Problems

Contamination

Contaminants, such as oil or "process" materials, may be introduced to a boiler through condensate return and may lead to deposit formation. Although process contaminants may occasionally form boiler deposits directly as a classic scale or sludge, those more commonly encountered cause deposit formation either by binding sludges or by charring on heat transfer surfaces. When the contaminant is an oily substance, the problem may be minimized by maintaining high concentrations of hydrate alkalinity and by using an organic dispersant. The best solution is to avoid the problem completely either through external treatment of condensate to remove contaminants, or by monitoring condensate and discarding it if contaminated.

Boiling Out

When a boiler contains oily or greasy deposits, a "boil out" usually is recommended. Such a procedure is also necessary on new boilers that usually contain protective greases to inhibit corrosion prior to use. In addition, dirt, debris, and mill scale must be removed.

Boil out is usually accomplished using an alkaline material, such as trisodium phosphate, sodium hydroxide, soda ash, or sodium silicate. A separate descaling with hydrochloric, or other acid, following degreasing may be necessary either for removal of operational deposits or for boilers with severe mill scale.

Mineral and Metal Deposits

The inefficiency and potential hazards created by a "dirty" boiler make cleaning advisable as soon as possible after discovery of the condition. Cleaning may be accomplished either on-stream or off-stream, as conditions warrant. On-stream cleaning often is recommended for low and medium pressure boilers that have a high ratio of hardness salts (Ca and Mg) to other deposit constituents and when deposit formation is uneven. One method of on-stream cleaning uses high concentrations of chelant, with polymers and organic sludge conditioners. Use of polymers to disturb the already formed deposit and to assist in breaking down deposit structure is important.

Organics are helpful in preventing redeposition. An on-stream cleaning program should be accompanied by heavy bottom blowdown, and is usually designed to take 30 to 90 days. Rapid on-stream deposit removal should be avoided to minimize the possibility of redeposition and/or plugging.

The choice of an off-stream cleaning program is determined by the nature of deposits to be removed. A relatively mild acid, such as sulfamic acid, may be used to remove deposits containing high concentrations of magnesium or calcium. More elaborate cleaning procedures may be needed when significant percentages of either metal oxides or silica are present. Chelants or acids usually are used for removal of metal oxides or mineral deposits; fluorides usually are used for silica.

Nonchemical Boiler Deposit Problems

No guarantee against boiler deposits exists even if a chemical treatment program is well designed and external treating equipment is operated effectively. Deposits frequently result from mechanical problems and are most often caused by operating conditions. Overloading, rapid load swings, improper burner alignment, and defects in boiler design can also cause deposit and corrosion problems.

Overheating

Local hot spots or overheating may be harmful to the boiler. Rapid steam formation at the heat transfer surface (film boiling) is a cause of boiler scale. When film boiling occurs, salts dissolved in the boiler water remain as a deposit.

Buildup of either baked-on sludge or organics may occur in local hot spots. These deposits cause the area to become more susceptible to failure from overheating. Rapid corrosion is another problem caused by overheating.

Flame impingement (when the flame actually touches boiler tubes) constitutes one of the prime causes for local overheating in a boiler. This condition is usually due to misalignment of burners and can be detected by adding a copper salt to the combustion air to give the flame a detectable blue-green color.

Overfiring of the boiler causes hot spots. If the boiler is operated above designed capacity, circulation of boiler water may be insufficient to cool heat transfer surfaces. Resultant overheating and subsequent failure may occur in areas of poor circulation.

All circulation problems are not related to overfiring. Inadequate circulation may develop in floor tubes, roof tubes, or riser tubes adjacent to downcomers. In horizontal tubes, water may not wet the top portion of the tube; in vertical tubes, poor circulation may cause inadequate flushing of heat transfer surfaces. These problems often result either from low loading of the boiler or from boiler design flaws. Tube failure resulting from overheating also can be caused by starvation, a condition where water supplied to an area of a tube in an operating boiler is inadequate. Starvation can result from deposits either in the lower bend of a screen tube or in the lower header of a wall tube.

Intermittent Use

Boilers used intermittently are subject to deposit problems. When a boiler is off-line, suspended solids settle and deposit on metal surfaces. In time, these deposits harden and bake on when the boiler is brought back on line. Deposits can develop during intermittent boiler use even with a water treatment program. Deposits are usually multilayered and may become quite thick.

Removal from Service

There is an advantage to removing all the sludge possible before a boiler is taken out of service. This can be accomplished by frequent bottom blowdown, for a period of one to three days, before the boiler is secured. Blowdown should be accompanied by increased dosages of sludge conditioner to keep solids in suspension.

Once removed from service, the boiler should not be drained until allowed to cool to atmospheric pressure. Draining a hot boiler can cause suspended solids to bake onto hot metal. After the boiler is drained, it is usually beneficial to rinse with high pressure water.

Afterboiler Section

The "afterboiler" section includes superheater, turbines, steam distribution system and equipment, and condensate lines. Deposits may form in these areas as a result of carryover, either raw water or process contamination, condenser leakage, inadequate return line treatment, and migration of corrosion products.

Deposits on turbine blades cause distortion of designed velocity and pressure profiles, imbalance, and corrosion. Deposits in the steam system and equipment reduce heat transfer and may cause malfunction of steam traps. Poorly functioning steam traps are a major cause of lost steam.

Deposits in condensate lines cause localized corrosion and generate additional deposit-forming corrosion products. Return line failures represent loss of valuable heat and high quality condensate, and may cause shutdowns. Whenever possible, condensate streams subject to contamination should be diverted to other uses or handled separately and treated. If returned, monitoring of pH, conductivity, or other specific analytical methods is advisable, with provisions for automatic condensate disposal. Composite sampling of various condensate streams may reveal a periodic contamination problem. In summary, afterboiler deposits usually are caused by corrosion and contamination. The best way to prevent afterboiler deposits is to minimize corrosion and contamination.

Preventing Condensate Corrosion

Condensate can be contaminated by evolution of gases from the boiler water and by air leakage into the steam-condensate system. The primary source of carbon dioxide in steam condensate systems is carbonate and bicarbonate alkalinity in boiler make-up water. Carbonates and bicarbonates, when subjected to boiler temperatures, undergo decomposition to form carbon dioxide.

Carbon dioxide dissolving in pure water reacts to form weakly ionized carbonic acid. Very little carbon dioxide is needed to sharply reduce the pH of pure water and substantially increase its corrosivity. In pure condensate at 140°F, for example, only 1 ppm dissolved carbon dioxide will decrease the pH from 6.5 to 5.5.

In the absence of oxygen, the acid condensate generally causes uniform attack, leaving a rather smooth surface where the iron has been dissolved away.

Dissolved oxygen is significantly more corrosive than carbon dioxide. Oxygen dissolved in pure water at 140°F, for example, has been found to be 6 to 10 times more corrosive to iron than molar equivalent concentrations of carbon dioxide. And, where both gases are present together in solution, the resultant corrosion rate may be 10 to 40 percent greater than the sum of the corrosion rates of the two gases acting separately.

The basic approaches to preventing condensate corrosion are:

- Eliminating (or minimizing) oxygen and carbon dioxide contamination
- Using chemical inhibitors to counteract corrosive conditions

Both approaches are necessary in a complete, successful program.

Feedwater oxygen can be eliminated by mechanical deaeration plus the use of an oxygen scavenger (catalyzed sodium sulfite or hydrazine). Chemical oxygen scavenging alone is also often effective. But even though feedwater oxygen can readily be reduced to zero by such means, it is difficult to completely eliminate air in-leakage into the steam condensate lines.

Various pretreatment methods can be used for reducing or eliminating makeup water carbonates and bicarbonates: lime softening gives some alkalinity decrease; special ion exchange processes (split-stream softening, dealkalization, demineralization, etc.) also reduce or remove alkalinity. Even with pretreatment to reduce potential carbon dioxide, chemical inhibitors are usually necessary to complete the job.

There are two basic types of chemicals used for preventing condensate corrosion: neutralizing inhibitors and filming inhibitors. Neutralizing inhibitors are volatile, alkaline chemicals that increase the condensate pH. They protect against carbon dioxide attack, but do not completely prevent oxygen corrosion. Filming inhibitors, properly applied, form a barrier between the metal and the condensate. They prevent both carbon dioxide and oxygen attack. The Food and Drug limitations on additives must be observed wherever steam or condensate comes in contact with food.

The most commonly used neutralizing inhibitors are amines like morpholine, cyclohexylamine, and diethylaminoethanol. The ability of each product mentioned to enter the condensate or water phase is indicated by its vapor-to-liquid distribution ratio. This ratio compares the concentration of amine in the vapor phase to the concentration in the water phase.

Product	Vapor-to-Liquid Distribution Ratio
Morpholine	.4 to 1
Cyclohexylamine	4.0 to 1
Diethylaminoethanol	7.0 to 1

In order to neutralize carbonic acid, the amine must be present in the water phase. The distribution ratio tells us which amines prefer the water phase and which prefer the steam phase. An amine such as morpholine preferring the water phase will be present in the initially formed condensate where the percentage of steam is large compared to the condensate. On the other hand, cyclohexylamine tends to remain with the steam to enter the condensate as the temperatures decrease and the percentage of steam in relation to condensate decreases.

Because of their differing vapor-to-liquid distribution ratios, two or more such amines may be used together to provide highly effective neutralization programs for complex systems.

Neutralizing amines are fed to the feedwater, boiler, or steam header. They are controlled by returned condensate pH from samples taken at the beginning, middle, and end of a condensate system.

SPECIAL HIGH TEMPERATURE
HOT WATER (HTHW) CONSIDERATIONS

HTHW heaters commonly are used for space heating and process purposes. Water heaters are similar to boilers in appearance, except that water is kept under pressure to prevent boiling. Normally, HTHW systems are closed. Impurities enter the system in makeup water.

In addition to normal low pressure boiler treatment, the following points should be observed for HTHW systems:

- Eliminate all leakage.

- Use an oxygen scavenger, such as hydrazine, to reduce oxygen corrosion in the initial fill.

- Fill the system with condensate or demineralized water. If not available, use soft water.

- Provide an initial treatment to coat and protect metal surfaces and to fluidize any precipitated solids.

- Assure that expansion tanks are bladder or piston type so air does not contact the water, or else charge the volume above the water in the tank with nitrogen.

COOLING WATER SYSTEMS

There are three types of cooling water systems used at naval installations:

- Once-through systems
- Closed recirculating systems
- Open recirculating systems

Pretreatment

Pretreatment of cooling water systems is necessary to ensure maximum service life of heat exchange equipment. Pretreatment consists of precleaning and prefilming.

Precleaning removes accumulations of foreign material that could reduce heat transfer, restrict flow, and impair corrosion protection. This procedure prepares surfaces of cooling water systems for the prefilming phase of the pretreatment program. Precleaning of either the entire system or individual components should be conducted just prior to start-up. Immediately after precleaning, surfaces should be passivated to avoid initial high corrosion rates. Special precautions should be taken with new cooling towers that typically are built with wood pretreated with a copper-based salt. It is imperative to flush out new towers to leach out these compounds.

Prefilming is a procedure that promotes rapid formation of an inhibiting film over surfaces of either the entire cooling water system or critical exchanger, depending on the method of application. Economics, discharge limitations, and time requirements dictate whether prefilming is applied to the entire system or to individual heat exchangers. Prefilming minimizes initial corrosion that occurs at start-up and allows for the most efficient application during a continuing corrosion inhibitor program.

The function of prefilming is to permit rapid formation of a uniform impervious film to stifle the corrosion reaction immediately. When this film is established, continuous low treatment levels will maintain the film intact and will avoid accumulation of corrosion products. The continuous low treatment level heals slight breaks that may occur from minor variations in environment. Whenever any serious changes in environment occur that cause destruction of the film, corrosion products can accumulate before the film is reestablished by treatment. Under these conditions, normal control should be restored and a prefilming program should be conducted to rapidly passivate the system.

Prefilming of equipment in cooling water systems is recommended:

- for all new heat exchange equipment,
- whenever cooling water equipment is acid-cleaned,
- whenever low pH excursions occur,
- whenever serious process leakage occurs, and
- immediately following start-up after turn-arounds or inspections.

While it is imperative to use pretreatment programs for all steel systems, because of higher corrosion rates that occur, the principles of pretreatment have application to cooling water systems in general. The application of principles of pretreatment for rapid film formation, followed by use of normal treatment levels for film maintenance, will prove beneficial in minimizing corrosion of the steel parts. In general, improved heat transfer, longer service life, and less plant maintenance will be realized.

Once-Through Systems

Once-through cooling systems use cooling water once before discharging to waste. Since even small cooling systems operating on a once-through basis use relatively large amounts of cooling water, these systems are generally employed only where water at a suitably low temperature is available in large volumes and at low cost.

The usual source of once-through cooling water is wells, rivers and lakes, where the only cost involved is for pumping. Generally, the only external treatment applied to once-through water taken from rivers and lakes is rough screening to remove objects that can damage pumps and heat exchange equipment. Because evaporation is negligible in these systems, mineral characteristics of both influent and effluent water are virtually the same.

Scale Deposits

Scale deposits formed in cooling water lines and heat exchange equipment may be of several types: precipitation of calcium carbonate; accumulated corrosion products; mud, silt, and other debris. It is important to recognize the mechanism of formation for each of these types of deposits and the corrective methods required to prevent them. Temperature, rate of heat transfer, calcium, sulfate, magnesium, silica, alkalinity, dissolved solids, and pH of the water are all factors affecting scale formation. Calcium carbonate is a chief ingredient of scale found in many once-through cooling water systems.

Prevention of Calcium Carbonate Scale

Softening raw water used in once-through systems usually is not economically feasible because of the large volumes of water used. If a softening plant exists to supply softened water for other plant purposes, using the facility to soften reasonable volumes of once-through cooling water may be practical.

The usual method of inhibiting calcium carbonate scale is use of deposit control agents. These agents inhibit scale both satisfactorily and at acceptable cost.

The deposit control agent used must prevent crystal growth and scale formation by permitting scale-forming salts to exist in an over-saturated condition without precipitating from solution. Commonly used agents are polyphosphates, polyacrylates, and organic phosphonates. Blending of these agents can combine advantages to solve a particular problem.

Iron Deposits

Once-through cooling water systems using iron-rich well waters, may experience scale formation caused by the oxidation of ferrous iron as well water comes into contact with oxygen. In such instances, either sequestrants or dispersants, which are specific in activity on ferrous and ferric ions, can prevent precipitation of iron. Such agents as polyphosphates, organic phosphonates, and polyelectrolytes also are suitable for treating this problem.

Iron deposits may occur in systems that use iron-containing river water as a source. Iron may initially deposit on piping and heat exchange surfaces of a once-through cooling system as ferric hydroxide. Oxygen from cooling water will tend to oxidize a portion of this deposit to ferric oxide, resulting in a deposit rich in both iron oxide and silt.

Treatments using combinations of organic phosphonates, polyacrylates, and selected surface-active agents have been successful in removing deposits containing high amounts of both iron and silica and in preventing further deposits from forming. Normally, for economic reasons, treatment chemical is fed intermittently. Both quality of the water source and analyses of the deposits in a particular system dictate frequency and concentration of the intermittent feed.

Iron deposits also result from corrosion products. This can be caused either by general corrosion of the piping in the cooling water system or by galvanic corrosion where two dissimilar metals join.

In addition to iron deposition, general fouling from mud, silt, and miscellaneous organic debris in cooling water can be a problem with river water makeup. This is especially true during periods of upset conditions such as rainy weather and high run-off. To protect critical cooling systems from general fouling due to these contaminants, surface-active deposit control agents have been used, with combinations of polyacrylates and biocides fed intermittently. These materials decrease the tendency of particles to settle out of the water in low flow areas and to stick to surfaces of the piping and critical cooling surfaces by modifying surface characteristics of the contaminants.

Figure 5-5 illustrates loss in fuel efficiency that can result as vacuum falls due to paper-thin deposits on tubes of a large surface condenser. These thin deposits, on the cooling waterside, are the result of both general fouling and microbiological growth and can be controlled through application of deposit control agents and microbiocides.

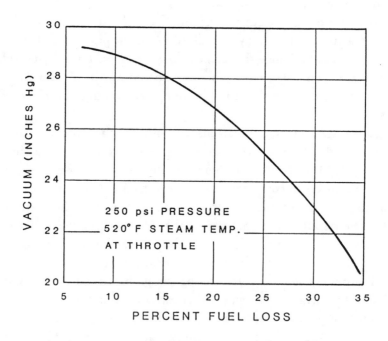

Figure 5-5. Fuel Efficiency Loss Resulting from
Condenser Deposits

Corrosion Control

Corrosion is defined as destruction of metal by chemical or electrochemical reaction with its environment. In cooling water systems, this metal destruction can cause several serious problems. Unchecked, it can lead to need for expensive replacement of equipment, and production losses from unscheduled downtime. An accumulation of corrosion products will reduce the capacity of piping and increase frictional resistance and pumping costs. Red water, caused by corrosion products, also is unsightly and may interfere with iron sensitive processes.

Closed Recirculating Systems

Removing heat from stationary engines and compressors requires attention from both plant operating personnel and water treatment consultants. In a closed recirculating cooling water system, Figure 5-6, water circulates in a closed cycle that is subjected to alternating cooling and heating without air contact. Heat, absorbed by the water in a closed system, normally is transferred by a water-to-water heat exchanger and then to the atmosphere or to some other heat sink.

Closed recirculating cooling water systems are suited to cooling engines and compressors. Diesel engines, in stationary and locomotive service, normally use radiator systems similar to a vehicle cooling system. Closed systems also are used in chilled water systems of air conditioners to transfer refrigerant cooling to air washer units where the air is chilled. The same system, in winter operation, can supply heat to air washers. Reliable industrial process temperature control is another area suited to a closed cooling water system.

Advantages of Closed Systems

Closed recirculating systems provide better control of cooling water temperatures through the heat-producing equipment and

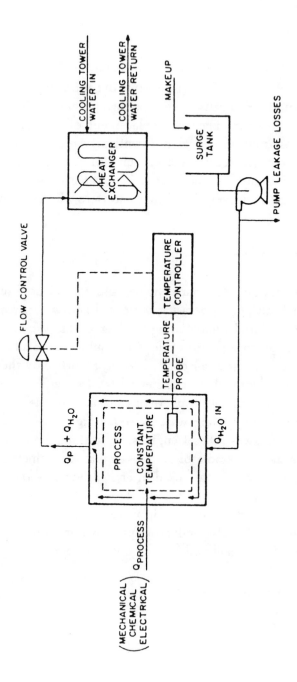

Figure 5-6. Typical Closed Cooling System

smaller makeup water requirements greatly simplify control of water-caused problems. Makeup water is needed only to replace losses from leakage at pump packings or when the system is drained for repairs. Little, if any, evaporation occurs.

By minimizing scale problems, these systems eliminate dangers such as cracked cylinders and broken heads. Moreover, closed systems are less susceptible to biological fouling by slime and algae, that results in scale deposits.

Closed systems also reduce corrosion problems because the recirculating water is not continuously saturated with oxygen, as in an open system. The only possible points of oxygen entry are at the surface of the surge tank or hot well. With the small amount of makeup water required, adequate treatment can virtually eliminate both corrosion and accumulation of corrosion products.

Scale Control

Closed systems that operate at relatively low temperatures require very little makeup water, and since no concentration of dissolved solids occurs, fairly hard makeup water can be used with little danger of scale formation. In diesel and gas engines, however, the high temperature of jacket water increases the tendency to deposit scale. Over long periods, additions of even small amounts of hard makeup water will cause a gradual buildup of scale in cylinders and cylinder heads. Where condensate is available, it is preferred as cooling water makeup. Where it is not available, softened makeup water should be used.

Corrosion Control

Untreated systems can suffer serious corrosion from oxygen pitting, galvanic action, and crevice attack. Closed cooling systems that frequently shut down will have water temperatures varying from 185°F to ambient. During shutdown periods, oxygen can enter water until its saturation limit is reached. When the system is returned to higher temperature operation, oxygen solubility drops allowing excess oxygen to attack metal surfaces.

Materials used in modern engines, compressors, and cooling systems include cast iron, steel, copper, copper alloys, aluminum, and zinc and tin alloy solders. Nonmetallic components such as either natural or synthetic rubber, asbestos, and carbon also are used. If bimetallic couples occur, galvanic corrosion may develop.

The three most reliable corrosion inhibitors, for closed cooling water systems, are chromate-based, silicate-based, and nitrite-based compounds. Chromate-based treatments generally are superior.

Open Recirculating Systems

Increasing interest in water conservation places new demands on water reuse programs. Industry is turning increasingly to recirculating systems. An open recirculating system with a cooling tower, spray pond, or evaporative condenser to dissipate heat, permits reuse of water and reduces the amount of makeup water needed. Such a system, however, intensifies potential for scaling, fouling, and corrosion.

Deposit Control

Selection of a deposit control program must be governed by an appraisal of interrelated factors that constitute the deposit problem. Assuring adequate control over blowdown rate may sometimes control deposits. Other situations require a sophisticated program that may involve at least two chemical deposit control agents and mechanical modifications, such as installation of a side-stream filter.

Deposit problems vary from tenacious films, barely visible to the naked eye, to plugging of heat exchanger tubes, and usually can be categorized as either scaling or general fouling.

Scale Formation

Scaling is precipitation of dense adherent material on heat exchange surfaces. Precipitation of scale-forming salts occurs when solubilities are exceeded because of either high concentration or increased temperature. Solubilities of most scale-forming salts in cooling water decrease with increasing temperature.

Passing circulating water through heat exchange equipment increases water temperature. This rise in temperature frequently is sufficient to cause precipitation of scaling material on heat transfer surfaces. The concentrating mechanism, in open recirculating systems, also is the cooling mechanism, that is, the evaporation of water when it passes through a spray pond or tower.

As pure water is evaporated, minerals in circulating water remain behind, resulting in water more concentrated than makeup water. The term that compares concentration in circulating water to makeup water is cycles of concentration. For example, circulating water that has twice the minerals concentration of makeup water has two cycles of concentration. Windage, or drift loss, fine droplets of water entrained in circulating air, and tower carryover, limit the degree of concentration. Makeup water must replace both windage and evaporation losses. Windage losses vary with the type of open recirculating system, but the following may be taken as typical:

Spray ponds .... 1.0 to 5.0 percent
Atmospheric towers .... 0.3 to 1.0 percent
Mechanical draft towers .... 0.1 to 0.3 percent

It should be noted that if sufficient biological control is not exercised in the cooling tower water, the growth and distribution of Legionnaire's bacteria through windage losses presents an extremely dangerous hazard to the general public in the vicinity.

Persons working on or in the vicinity of cooling towers and condensers that are being serviced could be exposed to Legionella and should take necessary safety precautions.

Guideline for protection of personnel:

1. Individuals that work in the mist of a cooling tower, either directly or indirectly in an aerosol-producing area of another facility, or within a high aerosol concentration of Legionnella, should wear personal protective equipment. Since inhalation is the entry route of concern, the equipment can be limited to respirators, with high efficiency or dust and mist cartridges. The cartridges should be discarded after each use.

2. Workers at high risk should be specially trained to use minimum water blasting to clean cooling towers and to avoid being downwind of an evaporative condenser or a cooling tower.

3. Water disposal from potential Legionella infested units should be done so that aerosols are not created.

4. New facilities should be located so that fresh air intakes of air conditioning units are not downwind from a cooling tower or evaporative condenser. Old installations should be modified to prevent such cross-contamination between units.

5. New cooling tower installation should have high efficiency drift eliminators, automatic water treatment, backflow devices to prevent cross-contamination, and sumps should be located so that backflow water contamination will be prevented.

Calcium carbonate scale is a potential problem in recirculating systems as it is in once-through systems. In addition, both scales of calcium and magnesium silicate and of calcium sulfate must be dealt with in recirculating systems.

When the makeup water illustrated in Table 5-3 is cycled two times, the circulating water remains nonscale-forming. When the same makeup water is cycled three times, scaling conditions are indicated. By limiting circulating water cycles and by adding acid, scale formation can be prevented. Scale inhibitors also are used to minimize formation.

Table 5-3. Circulating Water Analyses

	Makeup	Circulating Water at 2.0 Cycles	Circulating Water at 3.0 Cycles
Total Hardness as CaCO₃	60	120	180
Calcium as CaCO₃	50	100	150
Magnesium as CaCO₃........	10	20	30
P Alkalinity as CaCO₃	0	0	10
M Alkalinity as CaCO₃	50	100	150
Sulfate as SO₄...............	40	80	120
Chloride as Cl...............	10	20	30
Silica as SiO₂	5	10	15
		Above values expressed in ppm	
pH........................	7.0	7.3	8.3
pHₛ (140F)	7.8	7.3	6.8
Saturation Index	−0.8	0.0	+1.5
Stability Index	8.6	7.3	5.3
Interpretation	Non-scale-forming	Non-scale-forming	Scale-forming

Source: Betz Laboratories

Control of Cycles

Limiting cycles of concentration is one of the principal means of eliminating scale formation. By limiting mineral concentration of circulating water, over-saturation can be prevented or held within the effective range of the treatment used. Natural windage loss may limit cycles of concentration sufficiently. Where it does not, windage loss must be supplemented with blowdown.

Blowdown removes a portion of the concentrated circulating water, which is replaced with fresh makeup water. This lowers the mineral concentration in the system. Both continuous and intermittent blowdown are used. Continuous blowdown, however, is preferred.

While blowdown is an effective way to limit both cycles of concentration and the scaling potential of circulating water, excessive rates of blowdown may not be tolerable. Depending on water quality, blowdown cannot always accomplish scale control. In many localities, supplies of fresh water are either limited or costly. Rather than increase blowdown, acid addition, which permits higher cycles of concentration, is an approach to prevent scale formation. In recent years, the use of ozone as a cooling water biocide has made it possible to raise cycles of concentration to as high as 40, resulting in great savings in chemical treatment of water.

Corrosion of Non-Ferrous Metals

Copper and its alloys frequently are used in cooling water heat exchangers. Copper alloys are relatively resistant to corrosion, but specific chemical and mechanical factors permit corrosive attack.

Among the more common types of attack are dezincification, impingement, erosion-corrosion, stress cracking, and fatigue cracking. Because many copper alloys are available, an alloy should be selected that meets requirements of the particular service condition.

Some chemical factors that contribute to corrosion of copper alloys are low pH, ammonia, cyanides, sulfide, and excessive chlorine residuals. When attack occurs, the copper content of circulating water generally increases and can adversely affect the steel metallurgy in the system. Copper can deposit on steel and create cell action that will seriously pit the steel. Both chromate and nonchromate programs can provide protection for copper and copper alloys.

Monitoring and Control

Cooling water systems should have means of monitoring corrosion. This may be in the form of either a corrosion coupon bypass or a continuous corrosion monitor. A test heat exchanger also should be available to ascertain fouling characteristics and corrosion rates in the system under conditions that simulate plant operations. Data obtained from a monitoring device can help establish alterations to an inhibitor treatment program.

Microbiological Control

Control of biological fouling is necessary both to avoid heat transfer losses and to minimize added corrosion load imposed by slime and algal growths.

Both once-through and open recirculating water systems are susceptible to fouling, but generally the problem is more severe in recirculating systems. Exposure of circulating water to sunlight in both cooling towers and spray ponds encourages the growth of algae. Ideal water temperatures, higher concentrations of nutrients, and greater incidence of airborne contamination increase the slime-forming potential in open recirculating systems.

Fouling trends can be followed by plotting heat transfer rates versus time. Heat transfer coefficients may be calculated manually or may be determined instantaneously with an electronic instrument that integrates temperature and flow measurements. The degree of fouling can be closely monitored with test heat exchangers and deposition testers.

Deposition testers also are used to obtain an assessment of the deposit potential of water.

Corrosion

Preventing corrosion in open recirculating water systems has received much research attention. Effective anti-corrosion measures have been developed that can be applied at reasonable cost. The expense of corrosion prevention is minor when compared to the cost of downtime, production losses, and equipment replacement. Corrosion damage can require complete replacement of heat exchangers, pumps, and lines.

Causes of Corrosion: Because of the nature of recirculating systems, corrosion problems are intensified. Contact of cooling water with air creates unique corrosion problems. Continual replenishing of oxygen in cooling water, as it passes over a cooling tower, or through a spray pond, is the primary cause of corrosion in open recirculating systems. Airborne contaminants, such as gases (sulfur dioxide, ammonia, hydrogen sulfide), and particulate matter (sand, dust, dirt, fly ash) scrubbed from the air by cooling water, cause higher corrosion rates.

Prevention of Corrosion: Control of corrosion in open recirculating systems primarily is achieved by maintaining small quantities of chemical additives (corrosion inhibitors) in the cooling water. The corrosion inhibitors most frequently used in cooling water systems are classified as passivators. These types of inhibitors passivate metal by encouraging a metal oxide, or other film, to form on metal surfaces. Combinations of chromate, polyphosphates, and zinc are the most commonly used inhibitors. Other corrosion inhibitors include nitrites, silicates, amines, and various other organic agents.

Control of operating parameters such as inhibitor levels, pH, and dissolved solids in recirculating water is essential to achieve consistent corrosion inhibition. Control is vital when nonchromate treatment programs are utilized. Use of automatic control equipment greatly increases reliability of any corrosion inhibitor program.

Acid Treatment

Inexpensive sulfuric acid commonly is used to treat circulating water. Sufficient acid should be used to reduce, but not to eliminate, circulating water alkalinity. Alkalinity is reduced to achieve both saturation and stability indices that indicate nonscaling conditions.

When acid is used, the acid feeding equipment should be as automatic as possible to minimize adjustment by operating personnel. In most cases, investment in automatic pH control of acid feed is advantageous.

Fouling

Fouling of heat exchanger tubes usually is defined as deposition of nonscale-forming materials such as:

- silt or iron suspended in makeup water,
- naturally occurring organics in makeup water,
- particulate matter scrubbed from the atmosphere,
- deposition of chemical additives due to poor control,
- organic contamination from process leaks, and
- migrating corrosion products.

Control of Fouling: Treatment with dispersants is economically feasible for a wide variety of fouling substances. Low molecular weight polymers are effective dispersants in industrial cooling water systems.

Another approach to controlling deposition is to apply high molecular weight polymers (flocculants) that promote formation of a light, fluffy floc. Floc is removed through blowdown and/or side-stream filters. One hazard of using this approach is subsidence in low flow equipment, such as in the shell-side of heat exchangers.

On-stream desludging generally is designed for temporary relief from fouling. An on-stream desludging program must be tailored to suit individual chemical and mechanical requirements of a particular system. A typical program is conducted over 120 consecutive hours with the dispersant at the recommended increased dosage and blowdown increased to lower cycles of concentration. At the conclusion of a cleaning program, a high concentration of either continuous corrosion inhibitor or a prefilming agent is introduced to passivate the system.

Difficulties Due to Biological Fouling: Biological fouling, in open recirculating cooling systems, is the result of excessive growth and development of lower forms of life: algae, fungi, and bacteria. In general, the principal difference between algae and fungi is that algae can manufacture self-nutrients whereas fungi cannot. Except for a few autotrophic types, bacteria do not manufacture food.

Because sunlight is necessary for growth and development of algae, if not controlled, abundant algae growth occurs both in spray ponds and in exposed parts of cooling towers. The principal types of biological growth that occur in nonexposed portions of cooling systems are referred to as slime. Slime is an accumulation of microorganisms and their excretions together with inorganic and/or organic debris embedded in the mass. Microorganisms usually found in these deposits are various bacteria, filamentous fungi, yeast, and occasional protozoa. Slime growths also may contain dead algae that have migrated from other areas and have become entrapped in the mass. Slime growths may occur in either illuminated or dark areas.

Slime deposits on process equipment can retard heat transfer creating a serious loss in efficiency. Biological fouling on metal surfaces can create differential oxygen concentration cells and result in serious pitting of metal surfaces.

Choice of Biological Control Programs: Biological, growth-control materials in use are biocides and biostats. Biocides kill organisms; biostats inhibit growth and reproduction.

Whether a biocidal or biostatic agent is used, a residual concentration must be maintained for sufficient contact time if the agent is to function. For biostats to be effective, the residual must be maintained continuously.

Economics normally dictate intermittent treatment in cooling systems. Treatment periods vary from 15 minutes to as much as four hours. The object is to achieve an effective level of available residual at the system effluent. Dosages and contact times needed to accomplish this must be determined by experience.

Intermittent or shock treatment allows organisms to multiply for a time, then treatment removes organisms from system surfaces. The period needed for growth and removal varies in every system, even where the water source is the same. If a steady treatment is maintained, there will be no chance for biological growth, especially the

Legionella bacteria. This alone is reason enough for a steady chemical feed. However, shot feed at fixed intervals may be cheaper.

Many variables influence the biological development process. Each system must be considered individually, and different programs must be used for different seasons of the year.

Selection of a suitable treatment program requires definition of the existing, or anticipated problem. Deposits should be analyzed in the laboratory, source determined, rate of formation established, and importance of elimination considered. Test heat exchangers and test coupons that may be used in operating systems to help evaluation are valuable because use is feasible without interrupting system operation. System inspection provides the best information on degree of fouling.

Cost is the primary criterion in selecting a treatment program. Some factors that affect cost are: chemicals, equipment used to apply chemicals, manpower to administer the program, possible deleterious effects on the system, and possible final effluent composition. Other considerations are safety and handling.

Oxidizing Biocides: Toxicants used for biological control normally fall into two groups—oxidizing and non-oxidizing biocides. Oxidizing biocides are those that, in addition to their disinfecting action, oxidize other compounds.

Chlorine has a low chemical cost, is toxic to most microorganisms and reacts quickly, even in low concentrations. The amount of chlorine required to control biological fouling in a system is governed by several factors:

- quality of makeup water to the towers,

- amount and type of contact with the atmosphere,

- nature and quantity of system contamination from such sources as process leaks,

- other treatment programs for the cooling water, and

- type and amount of biological contamination.

Chlorine is available in either liquid form or a compressed gas, in such neutralized forms as sodium hypochlorite solution, dry calcium hypochlorite, and in organic compounds called chlorine donors. The feeding method, whether continuous or intermittent, and the residual chlorine required to control the problem are individual to a specific system and vary from tower to tower in the same plant. To control slime and algae in either once-through or open-recirculating systems:

- sufficient chlorine must be fed to secure and maintain a killing residual in the treated water, and

- residual must be maintained in the system long enough to control microorganisms.

Residual chlorine is the total amount of free and combined chlorine remaining in water after chlorine demand has been satsified. The difference between the dosage and the residual is chlorine demand.

While continuous chlorination is the most effective method of control, it is also most expensive. Table 5-4 compares advantages and disadvantages of different chlorination programs.

Table 5-4. Comparison of Chlorination Programs

PROGRAM	REMARKS
Continuous Chlorination — Free Residual	Most effective Most costly Not always technically or economically feasible due to high chlorine demand.
Continuous Chlorination — Combined Residual	Less effective Less costly Inadequate for severe problems.
Intermittent Chlorination — Free Residual	Usually effective Less costly than continuous chlorination.
Intermittent Chlorination — Combined Residual	Least effective Least costly.

Nonoxidizing Biocides: Chlorinated phenols are the nonoxidizing biocides most widely used in recirculating cooling water systems. These biocides are sodium salts of trichlorophenol and pentachlorophenol. Sodium pentachlorophenate, either alone or combined in commercial formulations with sodium trichlorophenate, probably is the most widely used chlorinated phenol. Sodium pentachlorophenate is a soluble and stable compound that does not react with most inorganic chemicals that may contaminate a cooling water system. Bromine compounds should be considered because of their nontoxic nature and fewer limitations on discharge.

With chlorinated phenols there are many different feeding programs that can be employed, ranging from continuous maintenance of a high concentration in the circulating water to intermittent addition of material in low concentration at infrequent intervals. As with chlorination, a slime control program must be fitted to individual system conditions.

It is desirable from an economic standpoint to employ intermittent feeding, if the problem can be controlled by this method. Unless the system has an unusually low retention time, there should be no marked difference in the inhibitory concentration required with continuous versus intermittent feed.

Evaluating Treatment Programs: Methods must be established to evaluate biological treatment programs. These include system inspections, chemical and microbiological examinations of the cooling water, variations in operating data, and effect on other treatment programs.

System inspections are usually limited to observations of open conduits of cooling towers. Units or portions of systems, should be inspected whenever possible. Evaluations can be made using a test heat exchanger that simulates equipment conditions. Corrosion test coupon racks, although valuable, do not include the effect of heat transfer.

Examining cooling water for total bacteria population counts is a common method for evaluating effectiveness of biocides. Population counts indicate what has occurred in the bulk water. They may not indicate the extent of biological growth on system surfaces. Other treatment programs affected by a microbiological control program include both corrosion and deposition control.

Cooling Tower Wood Deterioration

Wood is widely used in cooling tower construction. Wood deterioration often shortens the life of cooling towers from the anticipated 20 to 25 years to 10 years, or less. Repair and replacement costs may be excessive while cooling tower operation is inefficient.

Cooling tower wood can experience three main types of deterioration: chemical, biological, and physical. Rarely is one type present without the others; usually all three types occur simultaneously. When deterioration occurs, it is sometimes difficult to determine the type of attack responsible. Physical and chemical deterioration, which are more visible, renders wood more susceptible to biological attack. Delignification, the removal of the lignite component from the wood, may occur due to inappropriate chemical conditions. For the last few years, plastic and FRP towers have been finding favor.

Chemical Attack: Chemical deterioration of cooling tower wood commonly becomes evident in the form of delignification or cellulose destruction, resulting in loss of strength. Chemicals that cause delignification are oxidizing agents and alkaline materials. Attack is particularly severe when high chlorine residuals and high alkalinity concentrations occur simultaneously. Wood that has suffered chemical attack has a white, or bleached, appearance and its surface is fibrillated. Attack is restricted to wood surfaces and does not impair strength of unaffected areas. Wherever cascading water has a chance to wash away surface fibers, a severe thinning of the wood will occur. In serious cases, fibers plug screens and tubes and serve as focal points for corrosion.

Chemical attack occurs most frequently in the fill section and wetted portions of cooling towers where water contact is continuous.

It will also occur where alternately wet and dry conditions develop, such as on air intake louvers and other exterior surfaces, and in warm, moist areas of the plenum chamber of the tower.

Biological Attack: Biological attack of cooling tower wood is of two basic forms: soft, or surface rot, and internal decay.

Organisms that attack cooling tower wood use cellulose as food. Characteristically, attacked wood becomes dark, loses much of its strength, and also may become brash, soft, punky, cross-checked, or fibrillated.

Internal decay, normally restricted to the plenum area, cell partitions, access doors, drift eliminators, decks, fan housing, and supports, is a more severe form of biological attack. It is characterized externally by an apparent sound piece of wood that upon breaking, shows severe internal decay. Because decay is internal, it is difficult to detect in early stages.

Physical and Other Factors: One major physical factor affecting wood is temperature. High temperature affects wood adversely. Continuous exposure to high temperature produces gross changes in anatomical structure and accelerates loss of wood substance. These effects weaken the wood and predispose it to biological attack, particularly in the plenum areas of the tower.

Other factors influence the deterioration of tower wood. Areas adjacent to iron nails and other iron hardware usually deteriorate at an accelerated rate. These areas lose much of their strength and produce patches of wood that crumble easily. Slime and algal growths and depositions of dust and oil can aid growth and development of soft rot organisms.

Control of Wood Deterioration: The only effective method for protection of operating cooling towers is a preventive maintenance program. Preventive measures are relatively easy to accomplish for flooded sections of the tower where chemical and biological attack is limited to wood surfaces. Preventive measures for the nonflooded portions of the tower, where internal decay is the primary concern, are more difficult. The success of a program depends largely on adopting appropriate measures before infection reaches serious proportions.

Controlling chemical and biological surface attack of cooling tower wood in the flooded portions of the tower, is a water treatment problem that requires:

- use of nonoxidizing biocides to control slime and to prevent biological surface attack, and

- proper application and control to minimize chemical attack where chlorine is used.

Experience shows that where nonoxidizing biocides alone are used to control slime in tower systems, surface attack is minimal. In many cases, it is possible to use a variety of treatment programs that combine use of chlorine and a nonoxidizing biocide. Where a combination program is possible, chemical attack can be held to a minimum and biological attack can be controlled effectively.

Preventive maintenance programs for the nonflooded and plenum areas of the tower require:

- thorough periodic inspections,

- replacement of damaged wood, and

- periodic spraying of the plenum areas with fungicides.

Cooling towers should be inspected thoroughly at least once per year as part of a good preventive maintenance program. Several different wood preservatives are available. On the basis of incidence of internal decay, seven different pressure treatments have been rated in the following order of effectiveness:

- creosote,

- ammonia-copper-arsenite,

- acid copper chromate and copper naphthenate,

- chromated copper arsenate,

- pentachlorophenol,

- fluoride chromate arsenate phenol, and
- chlorinated paraffin.

Periodic spraying with an effective fungicide is an essential step in a preventive maintenance program. Plenum areas should be sprayed with fungicide to render the wood resistant to spread and growth of fungi. The object is to apply fungicide at a high enough concentration so the wood remains fungistatic until the next inspection and spraying.

Wood Examination: Since several types of wood deterioration occur in cooling towers, physical inspections and laboratory examinations of wood samples should be scheduled regularly. Service laboratories should be equipped with specialized equipment needed to determine data necessary to evaluate the condition of the tower wood.

Macroscopic examination of the wood will reveal the degree of grooving erosion and depth of surface attack. Macroscopic study also will reveal the surface structure of the wood, whether either chemical or biological decay is present, or incipient, and if so, to what extent.

Chlorine: Chlorine has been discussed at some length because until recently, it has been the most common effective treatment. Current limitations on chlorine concentration, however, have led to the development of alternative methods of treatment which are more effective, less expensive, and less offensive to the environment. These compounds include:

Bromine Chloride: This substance has a faster decay rate than chlorine and is more expensive, but is significantly more effective. Its major advantage, however, is that there is almost no residual toxicity in treated water.

Chlorine Dioxide: This material is produced by a simple generator at the point of use. Comments regarding bromine chloride apply equally to chlorine dioxide.

SAFE USE OF CHEMICALS

Concentrated chemicals used for plant water treatment should all be considered toxic, and breathing of vapors or dusts should be avoided. Personnel working with chlorine, acid, caustic, and other concentrated chemical feed systems should be properly trained and should wear rubber gloves, a rubber apron or suit, and a face shield. A quick-acting safety shower protected against freezing should be available wherever sulfuric acid is being handled. Chemical Safety Data Sheets published by the Manufacturing Chemists Association, 1825 Connecticut Avenue, N.W., Washington, D.C. 20009, provide detailed information for safe handling and use of sulfuric acid and strong caustic.

Concentrated sulfuric acid very gradually attacks steel so that hydrogen, a highly flammable gas, can be generated inside steel storage tanks or piping and can form explosive mixtures with air. Open lights, flames, and spark producing tools should not be permitted nearby. Level-sensing probes operating on electrical conductivity must be intrinsically safe. Steel or iron pipe should never be valved off at more than one point because hydrogen gas generated can develop dangerous pressures in a completely closed line or result in a blow-off when a valve is opened. Stainless steel or PVC piping preclude hydrogen generation.

The use of hydrazine as an oxygen scavenger is not without disadvantage. It can be absorbed into the body in toxic amounts through vapor inhalation. Furthermore, hydrazine has been classified as a suspect carcinogen by the Occupational Safety and Health Administration (OSHA) and the National Institute of Occupational Safety and Health (NIOSH). Standards for allowable exposure to it and its salts have been recommended at 1.0 ppm by OSHA and at 0.03 ppm by NIOSH. General practice, however, can expose workers to greater levels than proposed regulations would allow. Exposure to hydrazine vapor can be limited to acceptable levels through closed feed systems, but potential risks remain due to spills, leaks, system failures, and vapor diffusion. To eliminate the handling hazards of hydrazine and the disadvantages of sulfite, new products have been developed.

Special caution must be observed when using live steam in humidification systems. Problems have been experienced with additives such as morpholine and diethylaminoethanol (DEAE).

Special care should be observed where steam may be in direct contact with food. Under the Code of Federal Regulations (Title 21, Sect. 173.310), boiler-water additives may be safely used in the preparation of steam that will contact food under the following conditions:

1. The amount of additive is not in excess of that required for its functional purpose, and the amount of steam in contact with food does not exceed that required to produce the intended effect in or on the food.

2. The compounds are prepared from substances identified below and are subject to the limitations prescribed:

   Cyclohexylamine: Not to exceed 10 ppm in steam, and excluding use of such steam in contact with milk and milk products

   Diethylaminoethanol: Not to exceed 15 ppm in steam, and excluding use of such steam in contact with milk and milk products

   Hydrazine: Zero in steam

   Morpholine: Not to exceed 10 ppm in steam, and excluding use of such steam in contact with milk and milk products

   Octadecylamine: Not to exceed 3 ppm in steam, and excluding use of such steam in contact with milk and milk products.

3.   To ensure safe use of the additive, in addition to the other information required, labeling must include:

- The common or chemical names and the additives

- Adequate directions for use to ensure compliance with all provisions of the regulations.

6

Wastewater and Effluent Treatment

WASTES AND ENVIRONMENTAL CONCERNS

Waste products arise from both treatment and cleaning processes. These processes concentrate contaminants and generate potentially toxic waste products. Clarifiers produce sludge from both raw water and contaminants and chemical additives. Ion exchangers produce waste streams with high levels of dissolved solids, including regenerant chemicals. Table 6-1 summarizes wastes generated by various water treatment processes.

Waste Standards

In the United States, no specific standard applies only to water treatment wastes. The method generally used is to apply drinking water standards and multiply by 100. For ion exchange regenerants, this means that chlorides should not exceed 2500 ppm (3520 ppm calcium carbonate equivalent). Very likely sulfate would be controlled at about 2100 ppm because of a potential for calcium sulfate precipitation. Total solids limits are listed in Table 6-2 as 500 ppm; therefore 50,000 ppm TDS could be permitted in a waste stream.

247

Table 6-1. Water Generated by Treatment Processes

Treatment Process[1]	Character of Waste Produced	Waste Volume Percentage Flow	Example of Waste Weight[2] Dry Basis Pounds Solids/1,000 Gal. Processed
Rough screens	Large objects, debris	5-10	
Sedimentation	Sand, mud slurry		
Clarification	Usually acidic chemical sludge and settled matter	2-5	1.3
Cold lime softening	Alkaline chemical sludge and settled matter	2-5	1.7
Hot lime softening (+212°F)	Alkaline chemical sludge and settled matter	2-5	1.7
Aeration	Gaseous, possible air pollutant, such as hydrogen sulfide		
Filtration, gravity, or pressure	Sludge, suspended solids	2-5 (for packed bed units)	0.1-0.2
Adsorption, activated carbon for odors, tastes, color, organics	Exhausted carbon if not regenerated. Small amounts of carbon fines and other solids can appear in backwash. Carbon regeneration is separate process (usually thermal) in which air pollution problems must be met.		
Manganese zeolite, for iron removal	Iron oxide suspended solids	Similar to other filtration processes	
Miscellaneous, e.g. precoat, membrane, dual media filtration fine straining	As in other filters. Precoat waste includes precoat materials	1-5	0.1-0.2 (plus precoat materials when used)

Process			
Reverse osmosis[3]	Suspended and 90-99 percent of dissolved solids plus chemical pretreatment if required	10-50	1.0-2.0
Electrodialysis	Suspended and 80-95 percent of dissolved solids plus chemical pretreatment if required	10-75	1.5
Distillation	Concentrated dissolved and suspended solids		
Ion Exchange process[4] sodium cation	Dissolved calcium, magnesium and sodium chlorides	4-6	1.3
2-bed demineralization ..	Dissolved solids from feed plus regenerants	10-14	4-5
Mixed bed demineralization ...	Dissolved solids from feed plus regenerants	10-14	5
Internal processes	Chemicals are added directly into operating cycle. At least a portion of process steam containing added chemicals, dissolved and suspended solids from feed, and possibly contamination from process can be extracted from the cycle for disposal or treatment and recycle.		

[1] Processes are used alone or in various combinations, depending upon need.
[2] Amounts based on application of process to raw water shown in Table 5-3. These values do not necessarily apply when these processes are used in combinations.
[3] Feed must be relatively free of suspended matter.
[4] There are many variations. Listed here are a few of the most important.

Table 6-2. Public Health Service Drinking Water Standards

Characteristics	Suggested Limit Not To Be Exceeded	Cause for Rejection
Arsenic	0.001	0.05
Barium		1.0
Cadmium		0.01
Chloride*	250	—
Hexavalent Chromium		0.05
Copper	1	
Cyanide	0.01	0.2
Fluoride	0.7-1.2	1.4-2.4
Iron	0.3	—
Lead	—	0.05
Managanese	0.05	—
Nitrate*	45	—
Phenols	0.001	—
Selenium	—	0.01
Silver	—	0.05
Sulfate*	250	—
Total Dissolved Solids*	500	—
Zinc	5	—

*Items of most concern in spent regenerant wastes

Environmental Considerations

Programs to protect the environment and to conserve water resources have resulted in re-evaluations of once-through cooling systems by both designers and operating plant officials. Temperature increases of 10°F to 20°F can be expected in once-through system effluents. A resultant increase in temperature of a receiving body is dependent on its size and the amount of effluent. This is particularly true for large utility plants requiring extensive quantities of water. When thermal pollution problems exist, a cooling tower and/or spray pond may be added to cool water before discharge. Alternatively, an entire system may be converted to use an open-circulating cooling tower.

In other situations, restrictions on use of toxicants and/or products to control suspended solids reduce the potential for maintenance of cooling efficiency on a once-through basis. As water conservation and pollution control programs develop, the once-through cooling system design will become less attractive for construction of new systems. Existing systems will be subject to modification to meet new discharge limits.

The Environmental Protection Agency's Office of Pesticide Programs is empowered to regulate all aspects of pesticide application under the Federal Insecticide, Fungicide, and Rodenticide Act, as amended. A pesticide is defined as "any substance or mixture of substances intended for preventing, destroying, repelling or mitigating any pest, and any substance or mixture of substances intended for use as a plant regulator, defoliant or dessicant." The use of microbiocides for control of microbiological growth in cooling water systems falls under jurisdiction of the act.

In accordance with this regulation, microbiocide formulations must be registered with the EPA. In addition, other federal regulatory agencies such as the United States Department of Agriculture (USDA), Food and Drug Administration (FDA) and Occupational Safety and Health Administration (OSHA) can regulate chemical composition and use of various household and industrial microbiocides.

As effluent standards become more stringent, plant management must consider ways to reduce volume and to improve quality of plant discharge. Recycling plant wastewater provides an attractive answer especially when considering:

- reduction of discharge fees,
- reduction of supply costs,
- reduced treatment costs for cooling water,
- less capital required for future wastewater treatment facilities, and
- potential cutback of plant production caused by water shortages.

Economic advantages, coupled with government effluent restrictions, have increased the popularity of recycling wastewater into recirculating cooling systems. It is not simply a matter of changing plumbing and direction of water flow. Cooling system reliability depends on a comprehensive approach to controlling corrosion, scale, and most importantly, fouling of heat exchangers and piping when wastewater is recycled for cooling. Plant operations are dependent on the total effect and interaction of cooling system chemicals, dissolved and suspended solids, wastewater treatment processes, and effluent standards.

BOILER SYSTEMS

The three main sources of wastewater from boiler water systems are pretreatment wastes, boiler blowdown, and condensate losses. Losses of blowdown and condensate result in energy penalties through lost heat.

Chemical shortages, together with discharge volume limitations, prompt reuse of both ion exchange regenerants and rinse waters. Acid and caustic are often saved for reuse in the next regeneration step or for waste treatment pH control. Control is needed to prevent scaling.

In general, treatment wastes are precipitated nontoxic salts that can either be discharged in a dilute solution or be buried. Blowdown and condensate water are of better quality than raw water normally fed to a demineralizer system. Consequently, this water is recycled as shown in Figure 6-1.

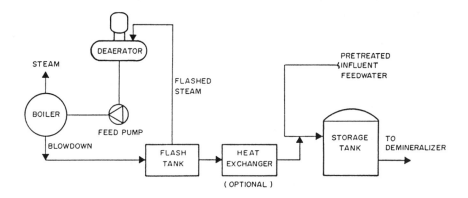

Figure 6-1. Boiler Blowdown Recovery

One potential complication may arise in the reuse of boiler blow-down: concentration of iron (and phosphates, if used for internal treatment) in the boiler. Either iron or phosphate-bearing sludge entering an ion-exchange unit can rapidly foul resin, impeding the demineralization process. This eventuality can be handled with carbon filters.

Another limitation is blowdown temperature imposed by either the demineralizer resin or the construction materials. Reducing the blowdown fraction usually can overcome this problem. As an alternative, water can be precooled in a heat exchanger. In this case, some heat may be recovered by heating feedwater.

Some plants use lime-soda softening of cooling tower blowdown to reduce calcium, magnesium, bicarbonate, and silica concentrations. By drawing blowdown from the warm return line, the plant can take advantage of improved hardness and silica removal at the higher temperatures.

When not using warm softeners, some plants combine blowdown with fresh makeup water in the primary lime-soda softener. This approach is generally more acceptable when the makeup water hardness is present as calcium or magnesium bicarbonate. Otherwise, high chloride and sulfate solids develop, creating a corrosive environment in the cooling system.

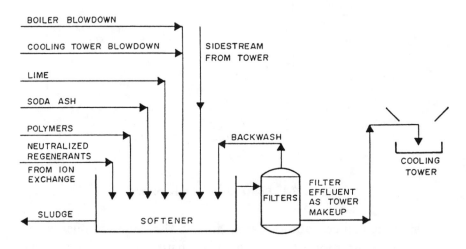

Figure 6-2. Schematic Diagram of Sidestream Treatment

COOLING TOWERS

Cooling water systems are the largest users of water, and consequently are the source of a major portion of plant wastewater. Water management calls for high cycles of concentration in cooling tower systems. Conservation can be partially achieved by eliminating indiscriminate blowdown to waste of recirculating cooling water used for pump gland cooling, continuous sample coolers, and pad washdown water. Such waste of cooling water increases the cost of makeup pretreatment, cooling water chemicals, and waste treatment.

Cascading, or reusing cooling water blowdown, can reduce wastewater and treatment costs under certain circumstances.

With proper pretreatment and effective recirculating water treatment and control at high cycles, cooling towers help conserve water.

Conservation is further aided by reuse of waste streams as makeup. Blowdown water usually must be processed to comply with environmental regulations before it is discharged.

Water is removed from cooling tower systems either continuously or intermittently, to prevent scaling and/or fouling. This is blowdown water. It contains suspended solids, concentrated salts, metal oxides and corrosion inhibitors, along with dispersants and microbiological agents.

Because chromate and zinc are prominent constituents of many effective cooling tower corrosion inhibitor programs, a discussion of blowdown treatment involves chromate removal or recycle. In some chromate treatment systems, blowdown water is prepared for discharge by either co-precipitation or exchange of other potential contaminants. Whatever the treatment, it is important to consider all aspects of environmental regulations for any particular blowdown stream.

Chromate and Zinc Removal

Environmental regulations severely limit chromate discharge, sometimes to a level as low as 0.05 ppm of chromium. This is a drinking water standard. To meet local regulations, the treatment process must be properly selected and effectively controlled.

Chromium functions as a corrosion inhibitor and, in this state of oxidation, is very soluble and stable. Chromate can be removed from cooling tower blowdown by: chemical reduction and precipitation, electrochemical reduction and precipitation, ion exchange, reverse osmosis.

Chemical Reduction

Several reducing agents are used to reduce chromate. The most common are sulfur dioxide, ferrous sulfate, sodium bisulfite, and sodium metabisulfite. Hydrogen sulfide, where it is available as a by-product waste material, occasionally serves as a chromate reducing agent.

After the chromate is reduced, the pH of the blowdown stream is raised to 8.5 with either lime or sodium hydroxide. At this pH, the solubility of chromium hydroxide essentially is zero and any zinc present is also precipitated.

Enough time must be allowed for chromium hydroxide to subside from the water, since this compound is a light flocculant substance that can take several hours to settle completely. This subsidence is usually carried out in either a lagoon or a conventional clarifier. If desired, a polymer can help clarify the blowdown water.

Chromate removal, whether with sulfur dioxide, ferrous sulfate, or other reactant, can be either a batch or a continuous process. The selection of a particular process depends on: volumetric rate to be treated, availability of chemicals (particularly waste chemicals), environmental considerations, economic considerations.

In the conventional chromate removal process, the sludge generated is both environmentally and economically important. Sludge is expensive to dewater, a procedure that must precede disposal. In some parts of the U.S., it is not permissible to land-fill chromium sludge.

Electrochemical Reduction

A chromate removal system based on the use of sacrificial iron electrodes is commercially available. Direct electrical current imposed on iron cathodes and anodes produces ferrous hydroxide.

The electrochemical process can tolerate a wide influent pH range of 6.0 to 9.0. Hydrogen is evolved from the cell, leaving an excess of hydroxyl ions that raises the pH of the water by as much as 0.5 unit.

The power requirement is five to seven kilowatt-hours per pound of chromate as Cr; iron electrode requirement is 3.5 pounds of iron per pound of chromate as Cr. About 85 percent of operating costs are for electrode replacement and 15 percent for power. The total cost of reduction is $0.05 to $0.10 per 1000 gallons of water, with the chromate at a level of 20 milligrams per liter (mg/l).

Ion Exchange Processes

Chromate in cooling tower blowdown exists as a relatively stable and soluble anion. It can be selectively removed in ionic form by exchange with chloride, sulfate, or hydroxide on special resins, and recovered for further use.

A schematic diagram of a typical ion exchange process is shown in Figure 6-3. Suspended solids can cause fouling and degradation of weak base resin. Since cooling tower water contains substantial amounts of suspended solids, the water must be filtered before reaching the ion exchange beds. Adjusting the pH to a level of 4.5 to 5.0 increases the holding capacity of the exchange resin. The pH can be adjusted by either adding acid or using a hydrogen cation unit.

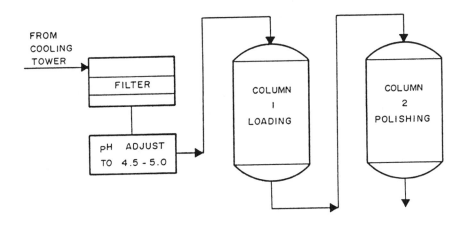

Figure 6-3. Typical Ion Exchange Process

The overall operational cost for a weak base process depends, to a great extent, on the life of the resin. Sulfuric acid and caustic soda costs range from $0.02 to $0.04 per 1000 gallons of blowdown water when chromate concentration is typically 20 mg/l. Resin replacement cost can be one to two times the chemical cost range, depending on influent water quality and operating conditions. The value of the recovered chromate, when reused as the cooling tower system corrosion inhibitor, can offset the operating costs.

A weak base anion resin system does not remove trivalent chromium or zinc. If these metals have to be extracted from the blowdown before discharge, a cation exchange unit can be superimposed on the system.

In addition to the value of the recovered chromate, one of the chief advantages of the ion exchange process is that sludge is not produced. Thus, fluid streams, such as rinse and backwash, can be returned to the cooling tower.

Blowdown Softening

A practical process for plants opting for chromate as an inhibitor and still meeting stringent environmental regulations is blowdown softening with lime and soda ash. The effluent water can be reused in the cooling system as makeup. A major advantage is recycling of the chromate to the tower system in the processed blowdown. A system of this type can approach zero blowdown where it is possible to limit losses to softener sludge and tower windage. Such tower losses are indicated by the symbol "E" in Figure 6-4.

Various process schemes are possible with blowdown or side-stream softening. The quality of the makeup water and the tolerance of the cooling system metallurgy for dissolved solids are important factors to consider.

High feed rates of lime and/or soda ash are required and the effluent can show higher hardness or alkalinity. One serious disadvantage is disposal of the volume of sludge from the softener. Any chromate occluded in the sludge can cause a disposal problem. In general, to determine whether blowdown softening is the best course to pursue, economics of a specific plant should be studied, along with appropriate pilot testing.

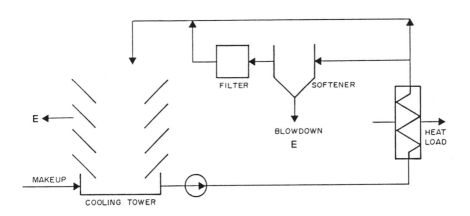

Figure 6-4. Chromate Recycle Blowdown Softening

Chromate

Chromate, long a favorite method of corrosin control, has fallen into disfavor because of the difficulty of removing it from waste streams prior to discharge to the environment. Alternate methods are now preferred, as good or better than chromate, and having less environmental impact. These are aminoethelyene phosphonates (AMP), sodium molybdate, and numerous nonheavy metal inhibirors. Considerable success has been obtained with combinations of materials, notably with AMP and molybdates, primarily because of the synergistic effect.

Insoluble starch xanthate (ISX), originally developed by the U.S. Department of Agriculture, has been proven to be one of the simplest and most effective ways to remove heavy metals from water. It can be used as an additive, or as a precoat material on a filter. The ISX exchanges sodium and magnesium ions upon contact with the heavy metal ions. The ISX containing the trapped ions is then removed from the water by precipitation or filtration.

7

Measurement of
Water Quality

Water is analyzed to determine the concentration of impurities that may contribute to corrosion, deposits, or other undesirable reactions. Analysis also measures effectiveness of corrective treatments being applied to the system. Analysis identifies the impurities and establishes a treatment strategy. Test results must be compared to prescribed limits and any deviation must be interpreted so that corrective measures can be taken. The type of tests, frequency, and control limits will vary, depending upon the system. There are two types of water quality analyses: sampling and laboratory analysis, and monitoring—either manual or automatic.

Sampling and laboratory analysis and monitoring programs are complementary. If extensive monitoring instrumentation is installed, laboratory analysis may be used only to check instrument readings. If little on-line instrumentation is used, a sampling and analysis program must be comprehensive. The balance between analysis and monitoring programs is determined by the size and type of system.

SAMPLING AND LABORATORY ANALYSIS

Diagnosis of either existing or potential water problems requires considerable information about the system. Thorough and accurate analyses of makeup water and water from operating systems, analyses and examinations of deposits and sections removed from operating systems, and information on the size, design characteristics, and operating conditions of the system are needed to make a decision about materials that must be either removed or added to the water to render it suitable.

Once the raw water supply has been chosen and a treatment system has been established, a control testing program insures adequate treatment rather than expensive overtreatment. A control testing schedule can vary from a simple color comparison for chromate in a closed cooling system to an elaborate determination of a dozen or more ions in the micrograms per liter range in the case of a supercritical boiler. The control program must be tailored for each individual case. For example, analysis for only chloride and phosphate concentrations may be necessary for treatment control in one type system, whereas in a very high-pressure boiler, alkalinity, chloride, copper, hydrazine or sulfite, iron, morpholine, pH, phosphate, sulfate, and analyses for other constituents may be required.

Table 7-1 lists the determinations most frequently performed to control the quality of water destined for steam generation, heating, cooling, and other manufacturing processes.

Methods of Analysis

The method of analysis for each constituent is as follows:

Suspended Solids: Gravimetric procedure must be used for quantitative analysis. In many cases visual inspection and descriptive terms—clear, hazy, cloudy, turbid—are used.

Soluble Solids: Gravimetric test is most accurate, but lengthy. Electrical-conductivity test gives quick, simple, and reasonably dependable results. Periodic gravimetric tests establish factors for converting electrical conductivity into ppm soluble solids.

Total Alkalinity: Titration test with standard acid uses methylorange indicator. End point is about 4.5 pH. Test measures alkaline

Table 7-1. Routine and Special Determination on Industrial Water Samples

Determination \ Applications	Boiler Water, Feedwater, or Condensate	Cooling Water, Recirculation (Open and Closed Systems) or Once-Through	Industrial Process Applications
Acidity	■	■	■
Alkalinity:			
Hydroxyl (OH)	■		■
Phenolphthalein (P)	■	■	■
Total: Methyl Orange or Mixed Indicator (M)	■	■	■
Ammonia	■		■
Boron		■	■
Calcium	■	■	■
Carbon Dioxide	■		■
Chloride	■	■	■
Chlorine, residual		■	■
Chromium, hexavalent	■	■	■
Color			■
Copper	■	■	■
Fluoride		■	■
Hardness	■	■	■
Hydrazine	■		
Iron	■	■	■
Lead			■
Magnesium	■	■	■
Moropholine	■		
Nickel	■		■
Nitrate	■	■	■
Nitrite	■		■
Octadecylamine	■		
Oil and Grease	■	■	■
Oxygen, dissolved	■		■
pH	■	■	■
Phosphate:			
Ortho	■	■	■
Poly	■	■	■
Residue, total (103 C)			
Filtrable	■	■	■
Nonfiltrable	■	■	■
Silica	■	■	■
Sodium	■	■	■
Specific Conductance	■	■	■
Sulfate		■	■
Sulfide		■	■
Sulfite	■	■	■
Tannin and Lignin			■
Turbidity			■
Zinc	■	■	■

hydroxides, carbonates, phosphates, and silicates. End point is vague unless titration is fast.

Phenolphthalein Alkalinity: Titration test with standard acid using phenolphthalein indicator gives sharp end point at about 8.3 pH. Measures all hydrates and half of the carbonates.

Hydroxide Alkalinity: Two titrametric methods are available. Simplest, using barium chloride, is accurate to 10 percent. Strontium-chloride method is accurate to about 2 percent, but lengthy.

pH: Colorimetric tests are accurate to about 0.1 pH. Electrometric methods using glass-electrode assemblies with meter are more sensitive and accurate. Faults with apparatus not readily discernible can introduce sizable errors. pH values can be continuously recorded.

Chloride: Two titrametric methods. Mohr's method uses silver nitrate with potassium chromate indicator—reasonably accurate for normal chloride-content waters. For low-chloride waters, more sensitive mercury-nitrate test is preferred.

Hardness: Versenate or EDTA titrametric method is extremely sensitive, accurate, relatively simple.

Sodium Sulfate: Gravimetric, photometric, and titrametric methods. Gravimetric procedure is too time-consuming for routine work and photometric tests are not accurate enough. Titrametric method using benzidine is sufficiently accurate for general use and is reasonably fast.

Sodium Nitrate: Colorimetric method uses either color comparator or electrophotometer.

Sodium Sulfite: One titrametric method uses N/40 iodine and N/40 sodium thiosulfate. Preferable method uses standard potassium iodate; a modification is available for low sodium-sulfite values of about 0 to 5 ppm.

Phosphate: Various color-comparison methods use either special color apparatus or an electrophotometer. Color comparators are relatively inexpensive, accurate, and easy to use; test is rapid and generally accepted.

Silica: Several colorimetric methods to choose from; gravimetric tests are too lengthy. Nessler tubes, color comparators, or electro-photometers can each measure color intensity. Silicomolybdate color test is most widely accepted and rapid, but requires careful technique. Use Chemetric's test ampoule and color comparator. Chemetrics test kits are available from water treatment firms, boiler supply houses and chemical suppliers. They are premeasured ampoules which provide a 5 second colorimetric analysis.

Dissolved Oxygen: Schwartz-Gurney modification of Winkler procedure is accepted method. Test demands considerable technical skill and is quite lengthy, with special equipment required. Indigo-carmine colorimetric method is new development. Short and simple to run, it is limited to range of 0 to 50 ppm dissolved oxygen.

Carbon Dioxide: Several titrametric methods are accurate only if carbon dioxide is the sole acidic constituent present. All are reasonably simple but require careful handling to get dependable results.

Ammonia: Direct nesslerization and measurement of developed color intensity in a colorimeter is rapid and accurate but dependable only on unbuffered water.

Hydrogen: Test of gas thermal conductivity is basic method. Results measure decomposition of steam in high-pressure boilers but require considerable skill to evaluate.

Steam and Vapor Purity: Gravimetric tests, throttling calorimeter readings, electrical-conductivity measurements, and tracer techniques are available. Conductivity measurements are most common, but specialized sampling apparatus is needed and sensitivity is limited at high purities. For steam purities of 1.0 ppm maximum solids this procedure is satisfactory. If more precise results are required, sodium-tracer techniques are used. If sodium salts predominate in the boiler water, flame photometer tests will detect sodium to 0.0002 ppm. This method is confined to limited spot-checking, in most cases, but can be continuously recorded.

Hydrazine: Colorimetric tests use comparator or electrophotometer. Special sampling technique is involved; test is lengthy.

Morpholine: Colorimetric method, a lengthy test using electro-photometer, requires technical skill.

Iron: A colorimetric method using ortho-phenanthroline and an electrophotometer is normal procedure.

Copper: Several colorimetric methods. Diethanolamine and carbonate test both require electrophotometer and skilled technique; a calibration curve must be prepared.

Note that all of the above analyses can now be adequately performed by automatic chemical or electronic analyzers. Also, the Chemetric ampoules now available can simply be broken off in the water sample, providing a color comparison test that is quick and cheap.

Sampling Guidelines

Location and design of sampling points need careful attention. Error will nullify all other efforts at boiler-water control, producing results that may be erratic and misleading. Sampling points usually should be located at input and output of important equipment in the cycle: heaters, boilers, and condensers. Figure 7-1 shows sample-point locations for several typical cycles. Frequency of sampling varies widely; the number of samples and tests at any point may be as high as ten to be sure of dependable data.

Continuously recorded results are justified at important points; at other points a single daily test will do. A careful review of the cycle dictates the frequency of tests. Continuous-recording meters should be considered. These instruments reveal erratic plant operation that may not be revealed by periodic sample testing. Automatic chemical analyzers are available as supplements to conductivity, pH, dissolved oxygen, and hydrogen recorders.

In gathering samples, care must be taken not to contaminate the water before testing. Design of the sampling point must take into account type of sample and tests to be run. The following sampling techniques should be used to reduce errors:

Water from Tanks: Water quality may be different at various depths in a tank. Take several samples at different depths.

Piping Runs: Water from pipelines can usually be tapped with simple sampling connections. Where water supplies combine or where

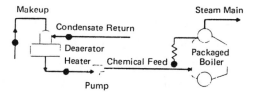

LOW-PRESSURE PACKAGED BOILER (With Minimum Treatment)

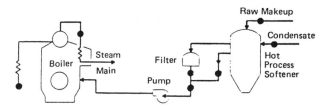

LOW-PRESSURE BOILER (With Hot-Process Softener)

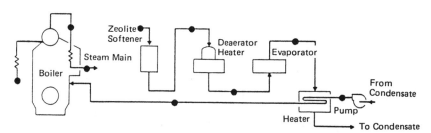

MEDIUM-PRESSURE BOILER (With Zeolite Softener and Evaporator)

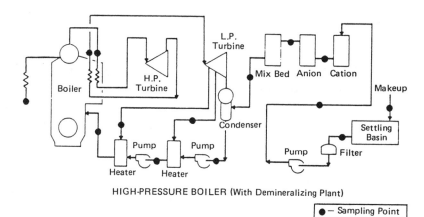

HIGH-PRESSURE BOILER (With Demineralizing Plant)

● — Sampling Point

Figure 7-1. Sampling-Point Locations for Typical Cycles

chemicals are introduced, the sampling point must be far enough downstream to allow complete mixing.

Boiler Drums: Water from boiler drums is usually sampled with a completely submerged perforated pipe extending the length of the drum. Avoid areas adjacent to feed inlet or chemical injection lines. Cool samples to prevent flashing.

Sample Handling: Sample lines must be thoroughly flushed two to five minutes before sampling. Time varies depending on length of line, pressure, flow, and sampling frequency. Use chemically resistant glass or plastic bottles—plastic for silica tests. Fill and rinse twice, then let bottle overflow long enough for three changes of water. Put stoppers on immediately. Where air contamination destroys test results, take sample underwater and stopper to avoid air bubbles. Test all samples as soon as possible.

Where several available methods will produce needed data, select a method that gives minimum required accuracy, and that can be run as quickly as possible at least cost.

Review the testing schedule regularly. Sometimes new test procedures are added, but tests no longer used in plant operation are not eliminated. Periodic reviews of testing programs are useful to eliminate unnecessary work.

Care of Test Equipment: Apparatus used for testing must be clean and in good repair at all times. Careful cleaning of equipment after use, and rinsing glassware prior to use helps assure accurate results. Containers of test reagents and indicators should be kept tightly closed when not in use, and care should be exercised to avoid contamination. Never pour unused reagent back into a storage container. It is good practice to discard reagents after six months and restock with fresh supplies. Test reagents, indicators, and other supplies should be obtained from a single source (usually a test equipment supplier or a treatment consultant) to insure compatibility and solution strength.

Location of the Test Station: Two important aspects of a water treatment program are the site selected for the water test station and the basic equipment provided for the test station. For most installations, the test station need not be elaborate; however, it should meet the following requirements:

- The area should be well lighted. Daylight type fluorescent lighting is preferred.
- The work surface should be sturdy and level and of sufficient size to provide uncrowded accommodation of test equipment, instruments, log books, and records.
- The location selected should not be in a high traffic area.
- A sink with running water is needed for cleaning equipment.
- Furnish a cabinet for storage of reagent stocks and test equipment.

STEAM SAMPLING

Steam purity, particularly steam for ships, is receiving high level attention. The following paragraphs outline a steam purity investigation program.

Analyzing Steam Purity

Steam purity test results must be carefully interpreted. Although the results of a steam purity study indicate that cycles of concentration in a boiler may be increased without causing carryover, cycles should be increased only if boiler cleanliness can be ensured. Often, when steam purity alone is used to determine cycles of concentration, the boiler is found to contain deposits. In such situations, the fuel saved by reduced boiler blowdown is offset by the fuel lost through the insulating effect of boiler deposits.

There are three aspects of steam purity measurement:

- accurate steam sampling and collection,
- accurate determination of the purity, and
- valid interpretation of the data.

Steam Sampling and Collection

Years of research conducted on sampling nozzle design and location have resulted in the nozzle designs specified by ASTM* and ASME.* Strict adherence to sampling material specifications is mandatory.

After selecting a nozzle location, the next problem is to extract the sample correctly. To do this, the rate of sample extraction must be calculated to provide a linear velocity for the sample entering the nozzle equal to the medium passing the nozzle. This method is iso-kinetic sampling. Sampling rates above or below a 1:1 velocity ratio cause distorted stream lines and nonrepresentative sampling. The results may be 50 to 100 percent higher or lower than actual. Steam flow, pressure, temperature, and steam line and nozzle diameters must be considered. Sampling rates must change with significant changes in load. A sample flow rate of 500 ml per minute with normal steam flow can be used in sizing the holes in the nozzle, the sample line, and the cooling coil.

The sampling line itself must not contaminate the sample and must be of sufficient strength for the service. Line size should be as small as possible to minimize retention time between the nozzle and the final sample-collection point. The line should slope downward in the direction of flow. The cooling coils should have sufficient surface to allow adequate cooling and must be capable of cooling the sample under full pressure. If samples are to be collected in bottles for later analysis, thesample should be at 25° to 30° Celsius to prevent formation of a vacuum in the sample bottle, which could draw in contaminants. Where superheated steam is sampled, some of the condensed, cooled steam must be recycled to desuperheat the steam as close to the sampling nozzle as possible. The bottles used for sampling must be specially prepared to assure the absence of impurities.

*American Society of Testing and Materials (ASTM); American Society of Mechanical Engineers (ASME).

Analyzing the Sample

The method selected is determined by sensitivity and accuracy required. Basically, there are five methods of measuring steam purity:

- determination of solids by wet gravimetric analysis of large (4 liters, or larger) samples,
- ion exchange,
- conductivity,
- sodium tracer (flame photometry or specific ion electrode),
- radioactive tracer, and
- determination of carryover by use of a throttling calorimeter.

Wet gravimetric analysis is cumbersome, inaccurate, and too time-consuming to be practical. Radioactive tracer method provides optimum accuracy but is very expensive and too difficult to apply. Ion exchange is useful only for determining average carryover during periods of several weeks and often does not distinguish between volatile amines, ammonia, or carbon dioxide. This method does allow for the measurement of low levels of all contaminants. The sodium tracer technique provides precise measurements, with sensitivity as low as 0.1 ppb.

The most commonly used measurement method is conductivity. For accurate, meaningful results, the sample must be stripped of gases including volatile amines, ammonia, and carbon dioxide. The Larson-Lane Steam Analyzer, shown schematically in Figure 7-2, is the most popular instrument for conductivity measurement. Electronic analysis of condensed steam samples is now available. The most common indicator is the chloride in the sample.

The foregoing comments apply to on-line sampling-analyzing techniques. Studies can be conducted using the "bottle study" method. Condensed steam samples are collected in small specially prepared (triple acid washed) containers for later laboratory analysis. Although the bottle study method is time-consuming and produces intermittent data, it allows impurities other than sodium to be measured. For example, shore-to-ship steam, because of its potential use for blanketing ship boilers and ship feedwater generation, has specific steam purity requirements relative to pH, conductivity, hardness, and silica (Appendix A).

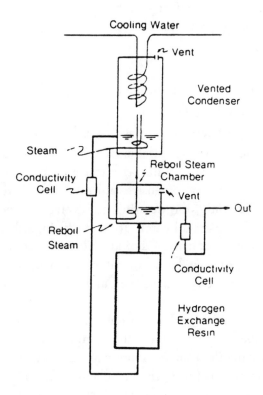

Figure 7-2. Schematic of the Larson-Lane Analyzer

Care must be taken to prevent contamination of both the sample and the sample bottle. An inverted funnel can be slipped over the tubing to cover the bottle during filling. The bottle should be opened only at the last moment before sampling, flushed 5 to 10 bottle volumes, filled to overflowing, and the cap should be rinsed in condensate before being replaced. Labeling with precise sampling time and source is critical for accurate comparison of the sample to steam flow, water level, and boiler water characteristics. Samples should be taken just before, frequently during, and just after every significant change in operation.

Regardless of the analytical method used, sample lines must be flushed 24 hours for a new line (otherwise, one hour) to remove contaminants. During actual sample collection, an isokinetic sample flow rate is required.

Conducting the Study

Before beginning tests, a study program should be established, abbreviated to suit a particular need or all-inclusive. A few variables to consider are:

- average steam load and steam load at several steady loads above and below average;
- steam load swings, large and small, fast and slow;
- water level, average as well as at several points above and below average;
- boiler water characteristics, particularly TDS;
- condensate contamination;
- variation along steam drum, to detect uneven steam release or faulty baffles; and
- internal treatment chemicals.

  Note: Soot blowing and bottom blowdown can disrupt a steam study and should be avoided while conducting tests.

Variables should be considered individually, and detailed records should be kept during the study. In addition to recording data pertinent to the variables, a fuel change, gas pressure drop, feedwater temperature change, and/or change in condensate return flow or source should be noted. Strip chart recorders for the steam analyzers should be used for on-line testing. Reference samples of the boiler water should be taken at a number of points during the test, especially before and after any change that would have a significant effect on boiler water chemistry. In addition to the regular analysis, sodium and gravimetric solids should be tested when using sodium tracer techniques.

ON-LINE MONITORING

Automatic Instrumentation

Until recently, most available automatic analysis equipment was designed for laboratory use. The problem with fully automatic measurement was a lack of equipment sufficiently reliable and rugged for field use. Automatic analytical instruments are now available for this severe service. Tests for specific conductance, pH, dissolved oxygen, phosphate, chromate, hardness, hydrazine, silica, and sulfite have been adapted to automatic control.

Conductivity

Dissolved solids, also called total dissolved solids (TDS), is a fundamental measure of water quality. Direct continuous measurement of TDS is difficult. Conductivity (in micromhos) is easily measured and is directly related to TDS. Many instruments are available that will measure specific conductance or conductivity.

Care must be exercised in selecting instruments. Often, the control system with a low initial cost will prove most expensive when the greater costs of maintenance, repalcement parts, and inaccurate control are considered. Instruments now exist that are highly reliable for water treatment applications. These instruments have features that include independent control and alarm settings, meter readout, recorder outputs, and automatic temperature compensation.

In both open cooling systems and boiler systems, conductivity is used for automatic blowdown control. Conductivity can be used to control an alarm or an automatic dump signal for many kinds of contamination in returned condensate. In most boiler systems, there is a satisfactory correlation between unneutralized specific conductance of the boiler water and blowdown demand.

Conductivity, combined with proper valve design and careful engineering for size, provides a reliable automatic system to control boiler blowdown. Depending on adjustment, the system automatically opens the blowdown valve for about one minute every 15 minutes. If conductivity is above set point, indicating the need for greater blowdown, the blowdown valve stays open until conductivity reaches an acceptable level.

pH

In cooling tower systems, pH is difficult to control because of sensitivity to small changes in acid requirement. pH measurement has been avoided as much as possible because of problems inherent in the measuring devices. pH electrodes have required continual cleaning and/or replenishment of reference solution. Electrical problems have occurred with the low-level signal, particularly in the humid conditions associated with water systems.

Electrode systems using a uniquely designed reference cell and a measuring electrode with integral preamplifier have overcome these problems. By coupling the electrode system with a pH transmitter-controller, a complete pH measurement and control system can be established.

Wet Analyzers

Certain colorimetric tests, such as for hardness and silica, have been adapted for continuous analysis. In a wet analyzer, reagents are added to the sample water and time is allowed for reaction and color development.

The sample-reagent solution is then discharged into a chamber where a photocell measures color intensity and corresponding concentration of a specific parameter. The measurement serves as a basis for either manual adjustment or remote-automatic regulation. Since the wet analyzer uses a transposed laboratory technique, it requires some care and attention.

Chemical Sensing Electrodes

The glass electrode used for pH measurement was the first to be developed to measure chemical concentration by an electrical effect. Other newer electrodes are sensitive to specific ions or to dissolved gases. Specific ion electrodes have a high degree of specificity for a single ion and can be used to measure calcium, chloride, fluoride, total hardness, cyanide, and others. Electrodes for measuring oxygen, ammonia, and sulfur dioxide are sensitive to the dissolved gas.

Typical Instrumentation

Figure 7-3 illustrates a recommended water quality monitoring instrumentation scheme for a high-pressure steam power plant. At most naval installations, automatic instrumentation is rare and, when installed, is poorly maintained.

The diagram shows the extent to which automatic analyzers and monitors may be employed. Such complexity is justified only in larger central plants.

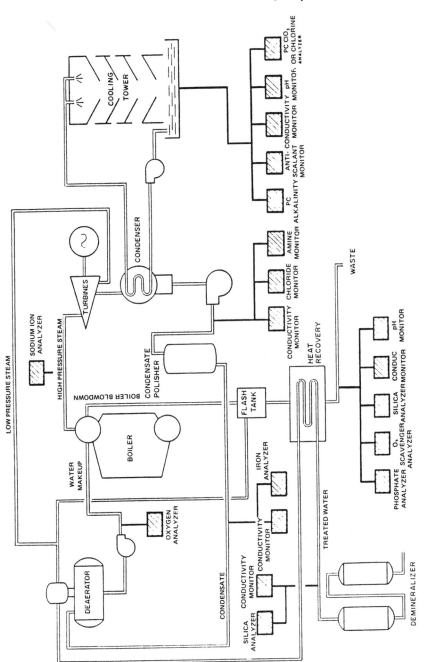

Figure 7-3. Analysis System for Large High-Pressure Boiler Plant

8

Corrosion

MECHANISMS OF CORROSION

Corrosion is an electrochemical process where a difference in electrical potential develops either between two dissimilar metals or between different parts involving the same metal. This voltage can be measured when a metal is electrically connected to a standard electrode. Electrical potential of a metal may be more or less than the standard; then the voltage is expressed as either "positive" or "negative." The difference in potential allows current to pass through the metal causing reactions at anodic and cathodic sites. These sites constitute a corrosion cell, as shown in Figure 8-1. The anode is the region of lower potential; the cathode is the region of higher potential. At the anode, metal ions go into solution. In general, the lower the potential of the anode, the greater the amount of metal dissolution, and the more serious the corrosion problem.

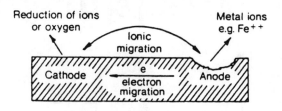

Figure 8-1. Simple Corrosion Cell

The extent of corrosion is a function of the capability of ions and electrons to travel through the water phase and to participate in chemical reactions. Water high in dissolved solids is more conductive and causes more severe corrosion problems. Seawaters generally are more corrosive than surface supplies.

Any metal immersed in water will develop a measurable potential. Metals of lower potential corrode more easily and extensively than those of higher potential. Theoretically, if two metals are coupled, the one of lower potential becomes the anode and corrodes.

Most corrosion occurs at the beginning of a metal's service life. Initially, metal dissolution is not impeded by a film of corrosion products. In time, a film will retard, or halt, corrosion. The degree that such a film can impede corrosion is a complex function of corrosion reactions, the structure of the deposit, and water velocity.

Cathodic Polarization

Polarization reduces the driving force of the corrosion reaction, and minimizes metal loss by changing the potential of either the anode or the cathode, or both, so that the difference in potential is reduced to a minimum.

Hydrogen bubbles off a cathode only when the cathode reaches a specific potential. The difference in potential between the cathode and a hydrogen electrode, at equilibrium in the same solution, is defined as hydrogen overvoltage. If available hydrogen produces more potential than the overvoltage, corrosion occurs. Overvoltage decreases with increasing temperature and surface roughness. The overvoltage needed for hydrogen evolution, for some common metals, is shown in Table 8-1.

Normally, the available hydrogen is insufficient to exceed overvoltage. In waters of low pH (acids with relatively high concentrations of hydrogen ions) overvoltage easily is overcome and corrosion occurs.

In natural waters, where pH levels are too high to overcome hydrogen overvoltage, the presence of dissolved oxygen usually controls the reaction. One corrosion control method involves governing the amount of oxygen available to the cathode surface. Oxygen is brought to the metal by convection through the bulk of the cooling water and then by diffusion through a thin laminar water film at the metal surface. If the amount of oxygen diffusion to the metal surface can be controlled, the corrosion reaction can be polarized. This is the mechanism of cathodic corrosion inhibitors. An impervious film is formed that prevents diffusion of oxygen to the cathode site. Another, more costly, way to remove oxygen uses mechanical de-aeration techniques most often employed in boiler plant operation. Mechanical deaeration is usually uneconomical for open cooling water systems. For ferrous-based materials, oxygen depolarization is a critical factor in almost all cooling water situations, since pH is maintained at levels where hydrogen evolution effects are minimal.

Table 8-1. Cathode Overvoltage of Hydrogen on Common Metals

Metal	Overvoltage in Volts
Platinum	0.12
Aluminum	0.19
Nickel	0.24
Iron	0.27
Silver	0.29
Copper	0.33
Artificial Graphite	0.35
Gold	0.36
Lead	0.42
Tin	0.49
Cadmium	0.50
Magnesium	0.59
Zinc	0.75

Closed systems are relatively easy to protect. Dissolved oxygen is rapidly consumed in formation of oxide films on metal surfaces. Since the system is closed, no further oxygen is available. The pH of the system is kept fairly alkaline to maximize hydrogen overvoltage. Therefore, the main cathodic reactions are under control. Compare this to an open cooling water system where heat is rejected by evaporation to the atmosphere. In this situation, the water is constantly resaturated with oxygen, with consequent availability to depolarize the corrosion cell. Open systems require more sophisticated corrosion inhibitor applications to maintain corrosion control.

Anodic Polarization

Anodic surfaces can be polarized by formation of a thin, impervious oxide layer. This film formation is accomplished by a mechanism known as chemisorption. Stainless steels form films naturally. Most metals, however, must be aided by addition of anodic corrosion inhibitors, such as chromate and nitrite.

Passivity

When corrosion reactions are completely polarized, the metal is said to be in a "passive state." At this point there is no difference in potential between the anode and cathode, and corrosion ceases. When polarization is disrupted in a passive metal at a point, a very active anodic site is set up, with resultant accelerated local corrosion, particularly if the metal is strongly anodically polarized. Note that deionized water is extremely corrosive when it contains oxygen, due to the lack of any elements to form the passive scale.

FACTORS AFFECTING CORROSION

Components of water, including dissolved and suspended matter, affect the method of corrosion control. The exact constituents of the water phase are unique to each water supply or application. Specific problems encountered in plant operations are similarly unique. Corrosion control requires consideration of each case separately. Often, a treatment program that works for one plant may fail for a neighboring installation. A summary of factors is provided in Table 8-2.

Table 8-2. Factors Affecting Corrosion

CHEMICAL

A. pH
    Acid Soluble Metals—oxides more soluble as pH decreases. Increased corrosion.
    Amphoteric Metals—oxides soluble at low or high pH. Protection favored at intermediate pH.
    Noble Metals—oxides insoluble at any pH. Inert to corrosion.

B. Dissolved Salts
    Chloride, Sulfate can penetrate passive metal oxide films and promote local attack.
    Calcium, Magnesium, Alkalinity may precipitate to form protective barrier deposits.

C. Dissolved Gases
    Carbon Dioxide—reduces pH and promotes acid attack.
    Oxygen—depolarizes corrosion reaction at cathode, oxygen deficient areas become anodic (differential aeration cell).
    Nitrogen—aggravates cavitation corrosion.
    Ammonia—selectively corrosive to copper based metals.
    Hydrogen Sulfide—promotes acid attack; forms deposits that promote galvanic corrosion.
    Chlorine—promotes acid attack, strips corrosion inhibitor films.

D. Suspended Solids
    Mud, sand, silt, clay, dirt, etc. settle to form deposits promoting differential aeration cell corrosion.

E. Microorganisms
    Promote acid attack, differential aeration cell corrosion, cathodic depolarization, galvanic corrosion.

PHYSICAL

A. Relative Areas
    In a galvanic couple, as ratio of cathodic to anode area increases, corrosion increases.

B. Temperature
    Increased temperature favors oxygen depolarization, lowers hydrogen overvoltage, increases corrosion.
    Higher temperature areas become anodic to other areas.
    Higher temperatures change metal potentials (e.g. reverse galvanizing).

C. Velocity
    High velocity promotes erosion corrosion, removes certain passivating corrosion products.
    Low velocity increases sedimentation and differential aeration cell corrosion, decreases amount of corrosion inhibitor reaching and passivating metal surfaces.

D. Heat Transfer
    Favors oxygen depolarization by "hot wall effects."
    Favors differential aeration cell formation by increasing precipitation and sedimentation of solids.

E. Metallurgy
    Surface flaws—cuts, nicks, scratches, etc. favor anodic site formation.
    Stress—internal stresses promote anodic site formation.
    Microstructure—metal inclusions, precipitation at grain boundaries, differing adjacent grains, etc. promote galvanic cell formation.

pH

Between pH 4.3 and pH 10.0 in natural, aerated water at ambient temperature, small changes in pH do not seriously affect the degree that steel corrodes. In the case of ferrous-based materials, the solution just outside the anode is mildly alkaline since it is generally saturated with ferrous hydroxide. Below pH 4.3, where there is free mineral acidity in the water, corrosion progresses rapidly. In this pH range, overvoltage plays a more significant role in the corrosion process.

The effect of pH on a metal is determined by the behavior of the metal oxide. If the oxide is soluble in acidic media, the metal will corrode rapidly in this environment. If the oxide will readily dissolve in alkaline media, there will be extensive corrosion in that pH range. Most metals fall into the first category.

Occasionally a metal oxide will dissolve in both acid and alkaline solutions; these metals are referred to as "amphoteric." These metals have their greatest stability, from a corrosion standpoint, at some intermediate pH range. Aluminum and zinc are amphoteric. Their corrosion rates are minimum at pH 6.5 and 11.5, respectively. Some metal oxides are insoluble at any pH. Their corrosion rate is independent of pH. "Noble" metals, those at the top of the galvanic series, behave in this way. Figure 8-2 illustrates this further.

A number of odd trends are evident in the behavior of iron. As the pH increases to 4.0, iron behavior is similar to an acid-soluble metal. Between pH 4.3 and 10.0, the corrosion rate is less influenced by pH, because oxygen depolarization is the principal factor determining corrosion. Any further pH increase reduces the corrosion rate, until a minimum value is attained at about pH 12. At pH 12 the corrosion rate again begins to rise with increasing pH.

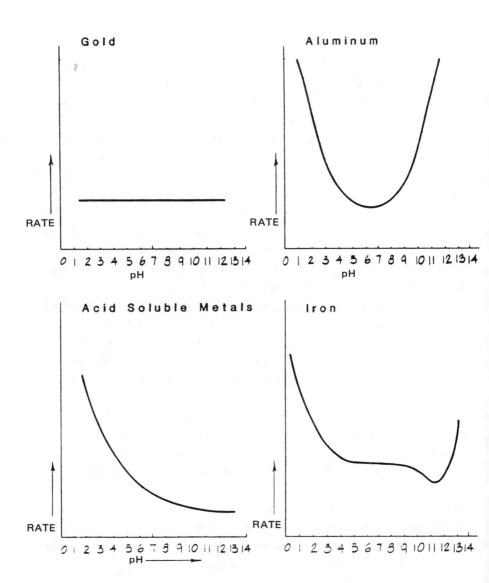

Figure 8-2. Variation in Corrosion Rate with pH Value

Courtesy of Drew Industrial Division, Ashland Chemical, Inc. Subsidiary of Ashland Oil, Inc.

Dissolved Salts

The corrosion rate for metals associated with natural, low-solids water, at normal temperature, accelerates as the concentration of dissolved salts in the water increases because of increased electrical conductivity. The corrosion rate may decrease in more heavily concentrated solutions, as a result of precipitation of dissolved salts and the formation of a barrier film that retards corrosion.

Dissolved Gases

A number of gases normally are found dissolved in water; these include carbon dioxide and oxygen. Other gases may be present as a result of contamination or other control programs. Examples of these are ammonia, hydrogen sulfide, and chlorine.

Carbon Dioxide

The solution of carbon dioxide into water decreases pH by the formation of carbonic acid, which promotes hydrogen evolution.

Oxygen

The amount of dissolved oxygen in water is directly related to its temperature, pressure, and surface area. Dissolved oxygen in water acts as a cathodic depolarizer, promoting corrosion.

A special case of oxygen corrosion involves unequal oxygen concentrations. A common manifestation of this is under-deposit corrosion. A porous deposit on a metal surface, whether from precipitated salts, suspended matter, or biological growth, almost always has an underlying oxygen-deficient environment. This leads to the formation of an active anodic site under the deposit and to severe localized corrosion.

Ammonia

Ammonia generally is introduced into water as a result of process contamination. It will selectively corrode copper.

Hydrogen Sulfide

Hydrogen sulfide, one of the most harmful gases that can enter a cooling water system, normally results from process contamination or from reduction of sulfate ions by sulfate-reducing bacteria. The gas promotes active corrosion in two ways: low pH attack and formation of iron sulfide, which is cathodic to iron and leads to galvanic corrosion.

Chlorine

Chlorine gas is the most commonly used toxicant for the control of microorganisms in cooling water systems. Chlorine reduces the pH of recirculating water and causes increased corrosion. On many metals it retards formation of certain protective corrosion inhibitor films.

Suspended Solids

Mud, sand, silt, clay, dirt, and other particles may enter a cooling water system either as airborne contamination or in the makeup water supply. Sedimentation of these materials takes place, porous deposits are easily formed, and differential aeration cells are quickly established. These can cause more corrosive damage than precipitated salts.

Microorganisms

Microbiological growth often presents special problems. Hydrogen is metabolized by many species, causing depolarization of the corrosion cell, similar to the action caused by dissolved oxygen. Anaerobic bacteria form differential aeration cells and accelerate local attack. Some species produce acidic compounds.

Physical Factors

A system's design and operating characteristics are often determining factors of the degree of corrosion. An understanding of these characteristics can assure optimum corrosion control.

Relative Metal Areas

The coupling of dissimilar metals produces a difference in potential, with the more negative the anode. The existence of an anode merely indicates that there is a potential for a corrosion current; it does not determine the rate of corrosion. This is governed by other factors, such as the relative areas of the cathode and anode. The rate of corrosion increases in proportion to the ratio of cathodic area to anodic area.

In the evaluation of a nickel-steel couple, for example, the galvanic series shows that nickel is cathodic to steel. In a tube-to-tube sheet joint in a seawater heat exchanger, using steel tubes and a nickel tube sheet, extensive corrosion of the steel tubes is expected. Reversing the situation results in far less corrosion of the steel, because the ratio of cathodic to anodic area is small.

The effect of relative areas also applies to a single metal. A small rupture in the oxide film of a passivated metal will cause extensive corrosion, because a small anode has been created in a large cathodic field.

The relative-area effects become more important as conductivity of the solution increases. In dilute solutions, only the cathodic area, immediately around the anode, is important, because the solution's conductivity is low. In brackish water, or seawater, relative area effects are extremely important.

Temperature

Corrosion increases with temperature. For example, in a domestic water system, a rise in temperature from 60°F to 176°F may increase corrosion as much as 400 percent. An increase in inhibitor is necessary to minimize this problem.

In an open system, corrosion rates increase with rising temperature. This trend continues until about 170°F, when the loss of dissolved oxygen exceeds the amount made available by diffusion, and a decrease in the corrosion rate occurs.

In a closed system, corrosion increases steadily as temperature rises since oxygen under pressure cannot escape.

Any temperature variation within one piece of metal causes the warmer portions to become anodic to the cooler areas, Figure 8-3. This problem is often experienced in fouled heat exchangers that are subject to uneven heat transfer.

Some metals, or alloys, change electrical potential as temperature changes. Zinc coating on galvanized steel becomes cathodic to the ferrous portion at about 150°F and does not provide corrosion protection.

Figure 8-3. Galvanic Cell Caused by Difference in Temperature

Velocity

There are two categories of water flow, laminar and turbulent. Laminar flow is low in velocity and may not be consistent across a metal surface. Turbulent flow is at a rate and distribution nearly uniform across a surface. Even in turbulent flow there is a thin, laminar film of water along the metal surface; the greater the turbulence, the thinner the laminar layer. Dissolved oxygen is distributed rapidly by turbulent mixing through the bulk of the fluid but encounters more difficulty in diffusing through a laminar zone to reach metal. As velocity (turbulence) increases, the depth of the laminar zone decreases and more oxygen reaches the metal surface. In addition, high velocity water removes what might become a passivating layer of corrosion product, producing the net effect of accelerated corrosion.

For water containing corrosion inhibitors there is a compensating factor. As oxygen diffusion to metal surface increases, so does inhibitor diffusion. Therefore, less corrosion inhibitor is required at higher velocities. Conversely, cooling water systems in a standby condition, or in low velocity areas (shell side water flow in a heat exchanger), are difficult to protect at normal inhibitor levels.

Heat Transfer

Heat transfer surfaces, such as in cooling water systems, are particularly difficult to protect. This is attributed to high metal skin temperatures. High temperatures can lead to "hot wall effects" that release oxygen from solution at the hot metal surface and promotes formation of a differential aeration cell.

Dissimilar Metals

Direct contact of dissimilar metals in a conductive solution may create a potential difference (with the metal of lower potential functioning as the anode) and result in formation of an active galvanic cell with corrosion of the anodic metal. If a galvanic cell is set up by the coupled metals, hydrogen will be reduced on the cathodic metal and absorbed on its surface.

The extent of corrosion is determined by many factors, the most important being: relative areas, solution conductivity, and polarization mechanisms. If the ratio of cathodic to anodic surface is low, galvanic corrosion is limited. Solutions of low conductivity also limit galvanic attack. Any polarizing condition in the system will decrease the potential difference between the metals and neutralize the corrosion reaction.

In the design of equipment, where galvanic corrosion may be a problem, dissimilar metals should be separated by a nonconductive substance such as a dielectric union, plastic sleeve, or insert.

Galvanic cells may be used to protect metal. In many situations, iron is protected by coupling to a metal such as magnesium. In the cell, iron becomes the protected cathode and magnesium the sacrificial anode.

Metallurgy

Metals are never absolutely flat, plane structures. All have surface scratches and crevices with potential for both electron loss and metal ion formation. Flawed areas become anodic to the bulk of the metal. A stressed metal normally sets up anodic sites at intergranular boundaries. Anodic site formation may result from a number of causes detectable under microscopic inspection. Inclusion of a non-homogeneous metal, or other metallic compound, in the grain structure results in formation of a small galvanic cell. Two adjacent grains of different densities might create a corrosion cell. Precipitation at metal grain boundaries causes a corrosion cell to form, especially if the precipitate is more noble than the metal itself.

An increase in metal purity does not guarantee that corrosion will decrease. Aluminum and iron serve as examples of contrasting behavior. Aluminum's resistance to corrosion increases as its purity increases. Resistance of iron remains the same as its purity increases. Pure iron is no more resistant to corrosion than either cast iron or steel. In the case of aluminum, corrosion protection depends upon the formation of oxide films, which is aided by increased purity. For iron, controlling factors are the corrosion reactions. Factors affecting corrosion are summarized in Table 8-2.

FORMS OF CORROSION

The following describes causes and characteristics of some common corrosion mechanisms affecting water-bearing equipment. Some are related to system design, some to composition of the water phase, and others to the pecularities of a particular metal.

Uniform Attack

If all corrosion results in even, uniform attack over an entire metal surface, the length of service and special design considerations could easily be determined in advance and equipment could be safely designed for the intended "life span." Unfortunately, this is not the case. Most corrosion occurs as localized attack at susceptible metal areas. It is more difficult to make predictions relating to design and service life under these conditions.

Pitting

Pitting is a type of localized attack commonly found, caused by formation of highly active, local anodic sites. Sites may be found at high temperature zones, at points of metallurgical defects, or at cuts, scratches, or crevices on metal surfaces.

Pitting is the most common cause of metal failure. One perforation may damage a critical heat exchanger sufficiently to disrupt an entire plant. Pitting corrosion is shown in Figure 8-4.

Figure 8-4. Small Corrosion Pits Distributed at Random

The depth of the pit is in direct proportion to the ratio of the large cathodic area to the small active anodic site. The seriousness of a pitting problem is often expressed in terms called the "pitting factor," the ratio of the pit depth to the average metal penetration.

Average metal penetration is often measured by recording the weight loss of corrosion test specimens over a period of time. Corrosion rates are expressed as mils per year (mpy). The higher the ratio of pit depth (in mils) to corrosion rate (mpy), the more serious the localized corrosion, and the greater the danger of metal failure.

Erosion

This form of corrosion is a class of attack associated with high velocity, and consists of two subclasses: impingement and cavitation.

Impingement

Turbulent flow, especially with water high in dissolved and suspended solids or dissolved and entrained gases, often damages passive oxide films, causing extensive local corrosion. The abrasiveness of the flowing water physically tears metal away from the structure and results in characteristic horseshoe-shaped pits.

In evaluating the effect of entrained gases, the size of the gas bubbles often determines the extent of pitting. Larger bubbles destroy the oxide layer, while smaller bubbles might bounce off the metal surface or actually replenish a passivating oxide film.

The most severe attack occurs at the inlet to heat exchanger tube bundles, points of restricted flow, sharp bends, and elbows. Although many materials are subject to impingement, the most serious problems occur in copper and copper alloys. In general cooling situations, flow through copper tubes in a heat exchanger should be limited to a velocity of five feet per second. This velocity should be reduced further if total solids or gas concentrations are high. Admiralty brasses, aluminum bronzes, and cupro-nickel alloys are more resistant to impingement. In noncritical heat transfer applications, impingement is often controlled by the addition of ferrous sulfate.

Cavitation

Cavitation occurs in high velocity, variable pressure situations and in water containing dissolved or entrained gas. Pockets of vapor form at low pressure regions, and as water flows to zones of higher pressure, the pockets collapse, causing shock pressures as great as several thousand psig. These high shock pressures physically destroy protective oxide films and dislodge metal grains from the surface.

Evidence of cavitation is often deep, circular pits, without tuberculation. Attack occurs in many areas, including the suction side of pump impellers, sharp bends in piping systems, and the discharge side of globe or gate valves. It may also occur in diesel engines, caused by the pressures created by piston slaps. High dissolved nitrogen concentrations have been found to aggravate the condition. Use of a suitable film-forming corrosion inhibitor can aid significantly in retarding attack.

Selective Leaching

Selective leaching occurs when one metal of an alloy is preferentially attacked and leached out of the alloy matrix. Many variations of this phenomenon exist but discussion is limited here to three commonly found forms: dezincification, graphitization, and dealuminification.

Dezincification

Leaching of zinc from brass leaves the metal with a weak porous copper structure, and a reddish, coppery color instead of the usual yellow color.

Certain conditions increase probability of attack, including oxygen differentials, low flows, high temperatures, acidic or alkaline media, and water with limited aeration. Primary areas of failure are under deposits in pipes, or in crevices formed by screw threads. The type of brass is also a factor in determining the extent of attack.

Graphitization

Cast iron and high temperature alloy steel may experience selective loss of iron, leaving the metal with a weak structure of graphite and iron oxides. Although the metal may appear sound, it can be penetrated with a sharp object. Graphite has a very low overvoltage and propagates attack by hydrogen evolution, usually occurring at grain boundaries initially and then spreading. It is favored by low pH, high dissolved solids, and acid contaminants like hydrogen sulfide gas.

Dealuminification

This form of leaching is associated with aluminum bronzes, and is usually found in seawater environments.

Under-Deposit Corrosion

Poor deposit control in cooling water systems can have far-reaching consequences. Heat transfer is reduced, causing possible undesirable process alterations or shutdown; corrosion is accelerated by formation of oxygen differential cells.

The water above a deposit contains dissolved oxygen, but the area below a deposit is oxygen deficient. A differential aeration cell is formed, with the oxygen deficient area becoming the anode. The result is extensive corrosion and high metal loss rate.

A self-perpetuating corrosion cycle is established. The metal lost through corrosion forms additional deposits that in turn create new differential aeration cells, and result in more corrosion.

Frequently the cycle induces rapid failure, despite the presence of a corrosion inhibitor. In this situation, rapid corrosion ensues with complete metal perforation not uncommon. It is essential to maintain a clean cooling water system to avoid costly downtime of critical process equipment.

Low-flow areas are particularly susceptible to this form of corrosion. Suspended materials settle in these areas, causing deposits and subsequent attack. This problem is magnified along heat exchange surfaces when deposition causes unequal heat transfer, which results in the creation of a corrosion cell, with cooler areas becoming cathodic to warmer ones.

Crevice Corrosion

Aggressive ions can concentrate in crevices; oxygen differentials can form and passivity is difficult to achieve. The degree of attack is proportional to the anodic crevice area compared to the surrounding cathodic area. Increased dosage of corrosion inhibitor is often needed to passivate these areas. Proper design provides the most practical solution. Rolling heat exchanger tubes over the tube sheet, for example, has prevented crevice corrosion failures at the tube-tube sheet joint.

Hydrogen embrittlement, a form of crevice corrosion, occurs mostly in older boilers with riveted seams. However, it may occur in any highly stressed surface in which there is a crack or a seam to concentrate an acidic OH ion.

Waterline Attack

There is, in any vessel, heat exchanger system, or distribution system not entirely filled, a three-phase region made up of air, water, and metal. The existence of this region results in simultaneous problems of differential aeration-cell corrosion and crevice corrosion. The oxygen depolarization reaction occurs easily in the region of the metal-water interface. This highly aerated region may become cathodic to other nearby metal regions. Cathodic inhibitors are generally employed to correct waterline attack. Care must be exercised because overfeed of zinc sulfate inhibitors has been known to accelerate the problem.

Cracking

There are two general forms of cracking corrosion: intergranular cracking and transgranular cracking. Attack is aided by high temperature, presence of high chloride concentrations, or other corrosion conditions.

CORROSION INHIBITORS

Corrosion inhibitors are classified as anodic, cathodic, or both, depending upon the corrosion reaction each controls. Inhibition usually results from one or more of three general mechanisms.

- The inhibitor molecule is adsorbed on the metal surface by the process of chemisorption, forming a thin protective film either by itself or in conjunction with metallic ions.
- The inhibitor causes formation of a protective film of metal oxides.
- The inhibitor reacts with a potentially corrosive substance in the water.

Choice of the proper inhibitor is determined by cooling system design parameters and by water composition. The type of metals in the system, stress conditions, cleanliness, and designed water velocity all affect inhibitor selection. Other factors to be considered include treatment levels required, pH, dissolved oxygen content, and salt and suspended matter composition.

Many of the factors discussed that affect the degree of corrosion in a system also affect choice and amount of inhibitor to be used. Lower levels are required in high velocity systems where laminar films along the metal surfaces are thin, permitting increased diffusion of inhibitor through the boundary layer, and better protection of the metal. Higher temperature systems usually require higher inhibitor concentrations to combat increased corrosion potential generated at these temperatures.

The salts in a water solution also affect inhibitor effectiveness. Hardness and alkalinity ions often aid inhibition by creating protective films. Chloride ions, however, increase the difficulty of passivating ferrous materials.

For corrosion control, each system must be considered as a separate entity. The program selected for a particular installation should reflect its unique design and associated water-phase characteristics, resulting in the best and most economically sound corrosion control program.

Anodic inhibitors build a thin protective film along the anode, increasing the potential at the anode and slowing the corrosion reaction. The film may eventually cover the entire metal surface. The appearance of the metal will be left unchanged to the naked eye.

Cathodic inhibitors generally are less effective than the anodic type. They often form a visible film along the cathode surface, which polarizes the metal by restricting the access of dissolved oxygen to the metal substrate. The film also acts to block hydrogen evolution sites and prevent the resultant depolarizing effect.

Cathodic inhibitors are considered safe because a low inhibitor concentration does not increase localized attack. Over-treatment can cause problems in some situations. For example, over-feeding of zinc sulfate, a common cathodic inhibitor, can increase waterline attack.

Following are descriptions of some of the more common corrosion inhibitors in current use, including synergistic blends of inhibitors prevalent in plant use.

Polyphosphates

Polyphosphate is a cathodic inhibitor that forms a durable polarizing film on the cathodic surfaces of most metals by an electrodeposition mechanism. As an additive to potable water supplies, polyphosphate stabilizes iron and eliminates "red water" problems.

Chromates

The single most effective inhibitor available today is probably the chromate or dichromate molecule. This anodic inhibitor forms a highly passive film of ferric and chromic oxides at the anode surface. The primary problems encountered with the use of chromates are environmental. Chromium, like other heavy metals, is known to be toxic to many forms of aquatic life. This has caused blowdown disposal problems. The National Pollution Discharge Elimination System restricts chromium discharge. State water quality standards, in many instances, mandate additional restrictions on discharge.

Zinc

Salts of zinc are the most commonly used cathodic inhibitors in cooling water systems. They rapidly form a film on the metal surface. Because the film is not very durable, zinc is usually not used alone but is found in many synergistic blends. Zinc presents toxicity problems to aquatic life, and its use has been restricted.

Nitrites

Nitrite formulations are used as metal passivators, especially in closed recirculating cooling water systems. Nitrites induce metal to form an impervious film for passivation. Sufficient nitrite is necessary for the film to remain intact. One disadvantage is that nitrite acts as a nutrient for various microorganisms; it either is oxidized to nitrate or reduced to ammonia by certain bacteriological species.

Silicates

Sodium silicates are used as corrosion inhibitors in potable water systems. They cause formation of negatively-charged colloidal particles that migrate to the anodic area and form a film. In open cooling water systems, silicates have been used with limited success. Mixtures of silicates and complex phosphates are used to improve control.

Benzoates

Salts of benzoate form a loosely held anodic film on ferrous materials. They are combined with nitrites to protect chilled water systems. Benzoates are highly effective in preventing waterline attack. There are two primary disadvantages to benzoates: they are costly and they are ineffective on nonferrous materials.

Lime

Alkalinity adjustment by the addition of lime is an inexpensive method of corrosion control that has been used for years in many municipal water supplies. Lime is added to adjust the Langelier Index from a corrosive level to just above zero on the scale-forming side. This technique is almost never used in open evaporating cooling water systems because of variations in system temperature and of effects on scale formation and corrosion.

Tannins and Lignins

Various types of tannins and lignins are used to control corrosion. Some compounds prevent cathodic depolarization by dissolved oxygen, others form an impervious film on metal surfaces, and others improve protection by modifying natural films.

Nitrates

Sodium nitrate has been used to protect solder and aluminum. It also increases corrosion inhibition of steel when used with other inhibitors such as nitrites. Its use in cooling water treatment primarily is in closed recirculating systems.

Surface Chelants

Surface chelants create, at the onset of corrosion at the anode, a monomolecular film on the metal surface. As metallic ions are formed, they react with the chelant molecules to form a tight layer on the metal surface, preventing further corrosion.

Phosphonates

These organic phosphorous compounds are similar in behavior to polyphosphates. Each can stabilize iron or hardness salts and form inhibitive films along metal surfaces. Care should be taken in using phosphonates on systems containing copper.

Molybdates

Certain molybdate salts are found equal, or superior, in corrosion inhibition to either chromate or nitrite salts used at higher concentrations. Sodium molybdate is an anodic inhibitor that forms a complex, passivating film. United States Public Health Bulletin No. 293 reports that the element has a "low order of toxicity." It has less environmental impact than either chromium or zinc.

Orthophosphates

These are anodic inhibitors. They are rarely used alone for corrosion control because of the danger of calcium sludge formation.

Aromatic Azoles

Mercaptobenzothiazole (MBT) is a specific corrosion inhibitor for copper and copper alloys and is chemisorbed on metal surfaces. It provides excellent protection at one or two ppm. The main disadvantages of MBT are chlorine demand, instability in the presence of oxidizing agents, and high cost.

Benzotriazole and polytriazole are superior to MBT as copper corrosion inhibitors. They are compatible with glycol antifreeze and are often used in closed systems constructed with copper materials.

Soluble Oils

When added to water, these anodic inhibitors form negatively charged particles that migrate to anodes and are precipitated with metal ions. The resultant oil film prevents further diffusion and controls the corrosion reaction. The oil film, however, can retard heat transfer.

Triethanolamine Phosphate

Used mostly in closed systems with or without glycol antifreezes, it offers more inhibition to steel as its concentration is increased, but less to copper. At several thousand ppm, both metals are protected.

Synergistic Blends

In plant operation, use of only one corrosion inhibitor in a cooling water system is rare. Usually, two or more inhibitors are blended to use the advantages of each and to minimize their respective limitations.

Frequently, anodic and cathodic inhibitors are combined to give better total metal protection (synergism). Many formulations blend two cathodic inhibitors to give additional polarization at the cathode. In less common situations two anodic inhibitors may be combined to give extra passivation.

A zinc-chromate formulation used at 40 to 50 ppm in a cooling tower system provides better protection than 200 ppm or more of chromate alone. Improved control often results from use of a blend in lieu of use of a single inhibitor. Although heavy metals still present problems, combining organic inhibitors into formulations minimizes the effects.

Pretreatment

Many corrosion inhibitors work best when applied at two to three times their normal dosage during the first several weeks of use. This procedure, called pretreatment, can result in improved corrosion control, particularly in new systems, because durable passivating films are quickly formed on metal surfaces. Pretreatment can be re-instituted to compensate for system upsets, pH excursions, corrosive contaminants, and prolonged low inhibitor control levels.

Inhibitor Performance

Industrial cooling water systems often are subjected to intermittent upsets that can affect a treatment program. Upsets can be caused by product-side contamination, high temperature, or low velocity.

Corrosion control in cooling water systems has reached a level of refinement where a corrosion rate of two mils per year or less is a realistic expectation. Table 8-3 shows the benefits of minimizing corrosion through use of inhibitors. The advent of blended inhibitors, especially those with the newer organic additives, provides an excellent source of protection. Combined with improved deposit control formulations, this has resulted in an effective total water treatment program.

Table 8-3. Typical Corrosion Rates Under Recirculating System Conditions

Treatment Program	*Dosage (ppm)	pH Range	Corrosion Rate (mpy)
Open System Inhibitors			
Chromate-zinc	50	6.5-7.0	0.7-1.9
Zinc-lignin	150	7.0-7.5	1.6-2.7
Zinc-phosphonate	75	7.0-7.5	1.8-2.6
Polyphosphate-phosphonate-polymer	100	7.0-7.5	1.7-2.4
Polyphosphate-zinc	50	7.0-7.5	2.2-3.4
Aromatic azole-phosphonate-lignin	150	8.0-8.5	2.6-3.6
Closed System Inhibitors			
Surface chelant	1000	7.0-7.5	0.1-1.3
Nitrite-borate-organic	2000	8.5-10.0	0.6-1.1
Sodium chromate	500	7.0-7.5	0.2-0.7
Untreated Control			50-100

*Dosage data is based on proprietary formulated products.

9

Ozone and Ultraviolet Treatments

OZONE[1]

Ozone, which has been used as a biocide for a hundred years, is finding increasing usefulness in primary water treatment, and more lately, as a treatment for cooling tower water. The Los Angeles Water Treatment (DWP) Plant, the largest in the world, has made studies prior to the installation of an ozone system.

Those studies demonstrated that ozone performs well as a primary disinfectant and in controlling the aesthetic parameters of water quality. They also found that ozone enhances the performances of other process steps and provides a number of water quality benefits not obtainable with other treatment methods, particularly in the area of turbidity removal.

It is generally considered that ozone acts as a flocculant aid or "microflocculant" by oxidizing organic substances which are often attached (adsorbed) to turbidity particles (or which constitute turbidity particles by themselves). Adsorbed organic substances can inhibit the coagulating action of conventional treatment chemicals which are necessary for effecting removal of inorganic turbidity particles

---

[1] Abstracted from the paper "Ozone, An Alternate Method of Treating Cooling Tower Water," by H. Banks Edwards, P.E. The paper was presented at the Cooling Tower Institute Annual Meeting, 1987.

during filtration. Oxidation of the organic substances essentially "cleans" the inorganic turbidity particles, exposing the charges on the particles and resulting in more efficient reactions with the coagulant chemicals.

Another explanation which may account for the microflocculation property of ozone lies in its ability to polarize organic substances with oxygen atoms. This provides charged sites on the inorganic turbidity particles, which were otherwise neutral or weakly charged because of their organic interferences. The new charged sites allow the particles to react effectively with the oppositely charged coagulant chemicals.

The microflocculating effect of ozone is significant because it allowed DWP to establish less conservative design criteria, which resulted in savings in the construction costs of other process facilities at the plant. Placement of the ozonation process ahead of other pretreatment processes provides the following design and operational advantages over conventional treatment alternatives.

In Plant Design:

- Increases filtration rates by 50 percent (from 9 gpm/sq ft to 13.5 gpm/sq ft), reducing by one-third the number of filters required.

- Decreases flocculation time by 50 percent (from 20 minutes to 10 minutes), reducing by one-half the number of flocculation compartments required; and

- Increases filter run times between backwash cycles, reducing the size of backwash facilities required.

In Plant Operations:

- Reduces chemical coagulant requirements by 33 percent.

- Reduces chlorination requirements by 25-50 percent, and

- Reduces filter backwash waste sludge (in proportion to reduction in coagulant usage).

In addition to providing microflocculation and primary disinfection, several other important water quality benefits are gained through ozonation. These include reductions in the levels of:

• Color

• Taste and odor

• Trihalomethane formation potential (by 50 percent), and

• Particulates remaining after filtration (detected by a particle counter rather than a turbidimeter).

Aside from this well-established type of use, it appears that ozone is the coming "miracle" treatment for cooling tower waters.

In 1970 M. Ogden published an article in *Industrial Water Engineering* discussing the merits of using ozone instead of chlorine for the treatment of cooling tower water. At about the same time, sales literature was published by Purogen, a Division of the Water Treatment Corporation, describing their process for using ozone to treat cooling tower water.

In 1976 the Water and Liquid Waste Treatment Advisory Task Force of NASA commissioned the Jet Propulsion Laboratory (JPL) of the California Institute of Technology to make a comparative study of the various methods of treating cooling tower water used by many of the NASA facilities. The study showed that all of the chemicals would perform as stated by the manufacturer if the instructions were rigidly followed. During this testing period, one tower used only ozone to treat cooling tower water. The ozone tests were so impressive that JPL had their commission extended to perform a series of tests using ozone exclusively in three cooling towers used in air conditioning service at the NASA facility located in Pasadena, CA.

The NASA/JPL project consisted of three cooling towers used in air conditioning service that would use ozone exclusively in the treatment of the cooling tower water. A fourth cooling tower would serve as a control tower using the standard chemical water treatment. The condensers of the three ozone-treated cooling towers were cleaned with inhibited hydrochloric acid and thoroughly inspected prior to the commencement of testing. During the test period these condensers were opened and inspected every three months to check the condition of the condenser tubing. One tower, with a 250-ton capacity was instrumented and used as the primary test tower. The other towers, a 75-ton and a 350-ton capacity, were used to check the results of the test tower. At the end of 21 months, all of the ozonated towers were found to be free of carbonate scale, but a thin film of inorganic sediment was found in the tubes.

The tests showed that corrosion rates for both copper and steel were about 50 percent of the rates when standard multichemicals were used. The copper coupon corrosion rate varied from 0.0 to 0.047 mils/year, but the Magna Corrator readings were below the limits of detectability. The steel coupon corrosion rate varied from 1.60 to 2.20 mils/year, while the Magna Corrator readings were less than 5.0 mils/year.

Table 9-1 (from the NASA/JPL testing) indicates the difference in constituent removal, cycles of concentration and ozone dosage of different cooling tower/evaporative condenser installations. The systems using ozone use tap water for makeup, no bleedoff, no other chemicals used and high cycles of concentration. Even though the cycles of concentration for ozonated cooling towers and evaporated condensers were much higher compared to the usual multichemical treated towers, the ozonated cooling towers/evaporative condensers removed a greater percentage of constituents under more severe conditions. The cycles of concentration in the ozonated cooling towers/ evaporative condensers ranged from 30 to 46 cycles compared to 2.3 to 10.5 cycles of concentration for multichemical treated towers. Despite these high levels of concentration, the water was clear and free of any color.

The water consumption of a cooling tower is figured as a percent of the water in circulation. For Evaporation the rate is 1.8 percent; for Drift Loss it is 0.1 percent; and for the Blowdown rate it is 0.9 percent for the cooling towers using the conventional multichemical water treatment system and the water circulation rate of 3 gpm per

Table 9-1. Constituent Removal from Cooling Tower Water and Ozone Dosage

Cooling Tower Systems	Cycles of Concentration	Constituent Removal, %						Ozone Dosage mg/l (ppm)		No Bleed-Off
		Ca	Mn	SiO$_2$	Alm	TDS	Thm	By-Pass	Circulation	
Ozonated										
NBC—Burbank, CA	46	73	27	73	96	23	99+	0.6	0.0004	
JPL—NASA	30	97	45	84	90	57	99+	0.5	0.003	
Bullocks, Sherman Oaks	30	46	34	70	55	36	99+	0.4	0.01	
Duke Univ. Med. Ctr.	44	46	34	70	54	19		0.3	0.02	
Chemical								Continuous Bleed-Off		
JPL—NASA	10.5	71	33	64	72	41		0.6 l/Joule/hr (2 gph/ton)		
Duke Univ. No. 1	2.3	4	4*	17	28*	39*	99			
Duke Univ. No. 2	2.4	8	4	17	21*	38*				

*Calculated

Data Source—Humphrey, Merrill

ton of refrigeration. For the NASA/Jet Propulsion Laboratory tests, a blowdown rate of 2 gph/ton or 1 percent was considered to be a more practical rate for the average cooling tower system.

Table 9-2 gives a cost comparison between the use of ozone (NASA/JPL tests) and the standard chemical treatment of cooling tower/evaporative condenser water. As shown, this savings amounts to $21.76/ton/year (1977 prices) with ozone treatment. More than 50 percent of the savings was the elimination of bleed-off of 2 gallons/ton/hour. It should be emphasized that the use of ozone for the treatment of the cooling tower/evaporative condenser water was profitable, energy efficient and environmentally responsible. Discharges, both water and air, from ozone-treated systems were in compliance with USEPA regulations.

In 1978 the Electric Power Research Institute (EPRI) commissioned Brown and Caldwell of Walnut Creek, CA to make a study of cooling towers and evaporative condenser systems used in air conditioning systems that used ozone in treating cooling tower and/or evaporative condenser water. The NASA/JPL was one of the systems studied. The result of the EPRI study is used in Table 9-1.

Brown & Caldwell reported that the corrosion rate of steel was reduced by 50 percent when ozone was used in lieu of chlorine for the treatment of cooling tower water. The towers reported on were at the Hoechst AG, West Germany and Messer Griesheim, GmbH, Dusseldorf, West Germany. Both of these companies are large chemical plants. They also stated that when comparing chlorinated versus ozonated circulated water systems, twice as many chlorinated hydrocarbons occurred in the chlorinated systems relative to the ozonated systems. Ozone destroyed tower algae, bacteria and fungi. The residual ozone prevented new fouling in the plants. After a one year period, the corrosion rate fell from 0.10 to 0.05 mm/yr for chlorine to 0.05 to 0.025 mm/yr for ozone; this was a 50 percent reduction. The fact that chlorine destroys the protective layer build-up by the inhibitor and ozone does not destroy this layer may explain the differences in the corrosion rates for steel.

Table 9-2. Cost of Chemicals vs. Ozone for Treating Cooling Tower Water

Tower No. 215 (JPL/NASA) 250 Tons	Chemical System		Ozone System	
	$/Joule/hr ($/Ton/yr)	Total	$/Joule/hr ($/Ton/yr)	Total
Chemicals				
Anti-Scale	0.24×10^{-6} (3.04)		0	
Anti-Corrosion	0.10×10^{-6} (2.48)		0	
Biocide	0.09×10^{-6} (1.24)		0	
Acid (pH)	0.02×10^{-6} (0.26)		0	
		(7.02)		0
Water				
Blow-Down*	0.87×10^{-6} (11.03)		0	
Basin Water Flush	0		0.02×10^{-6} (0.30)	
Acid Cleaning	0		0.02×10^{-6} (0.30)	
Water	0.04×10^{-6} (0.45)		0.01×10^{-6} (0.21)	
Acid	0.03×10^{-6} (0.32)		0	
		(11.60)		(0.81)
Electrical	0.01×10^{-6} (0.12)	(0.12)	0.11×10^{-6} (1.4)	(1.41)
Amortization (Equip.)	0.56×10^{-6} (7.04)	(7.04)	0.41×10^{-6} (5.36)	(5.36)
Labor (Main. & Mon.)	1.05×10^{-6} (13.28)	(13.28)	0.79×10^{-6} (9.93)	(9.93)
Grand Total	3.12×10^{-6} (39.26)	(39.26)	1.39×10^{-6} (17.50)	(17.50)

Savings — $39.28-$17.50-$21.76/Ton/Year (1978)
*Blow-Down - 2 gph/Ton   Data Source - Humphrey

In 1985 at the 46th Annual Meeting of the International Water Conference, a paper was presented entitled "The Investigation and Application of Ozone for Cooling Water Treatment" for cooling towers at the NASA Kennedy Space Center, Florida. This paper, together with the NASA/JPL report confirmed the benefits of ozone treatment of cooling tower water.

Following the completion of the review of the 1983 NASA/JPL study, NASA Kennedy Space Center (KSC) applied for funding to investigate ozone application to their cooling tower water systems. Due to the hot and humid conditions at KSC, it was determined to test to see if ozone would work to control microbial growth at KSC. In January 1984 EG&G, the Base Operations Contractor for KSC, was notified by NASA that funds were available for the study. After a review of the types of ozone generators, the most technically advanced unit available was selected.

The cooling tower system selected for ozone treatment recirculated 1,700 gpm, had a total water capacity of 6,000 gallons and was part of a 600-ton air-conditioning system. This system consisted of two water-cooled condensers. Both of the condensers were opened for inspection prior to commencing the tests. One was cleaned with wire brushes, while the second one was not cleaned. The testing began on August 1, 1984 and the final data of the testing was taken at the end of 12 months. The results of the testing are as follows:

1.   Bacteria count — before ozone — mean:
     3.0×1.06 CFU/ml; range: 2.1×1.04 to 6×10^7 CFU/ml.
     after ozone — mean: 2.5×1.02 CFU/ml; range: 2 to 3×1.02 CFU/ml.

2.   Turbidity — before ozone — 30 to 60 JTU
     after ozone — 5 to 10 JTU

3.   Conductivity — before ozone — 2,300 to 2,700 micromhos
     after ozone — 4,000 to 4,500 micromhos;
     after reaching 4,000 micromhos an accumulation of sludge and loose scale collected in the tower basin several inches deep.

4.   Cleanliness of the system — after a few weeks of ozone treatment, a noticeable change in tower appearance occurred. Microbial slime layers on the slats and fill surfaces began to turn gray and scale accumulation on the surfaces began to fall off.

5.   The ozone residual increased from 0.01–0.10 ppm to 0.10–0.30 ppm.

The conclusions reached as a result of the tests were:

a.   Tables 9-3 and 9-4 describe the source water quality characteristics and the monitoring schedule for the KSC towers.

Table 9-3. Make-Up Water Characteristics
Kennedy Space Center

Parameters	Average Concentration
pH	7.8 to 8.4
Conductivity	650-750 umho
Alkalinity, Phenolphthalein	0-6 ppm
Alkalinity, Methyl Orange	40-70 ppm
Calcium Hardness	60-90 ppm
Total Hardness	100-130 ppm
Chlorides as Cl_2	180-220 ppm

Data Source - Kennedy Space Center

Table 9-4. Parameters Monitored and Frequency
Kennedy Space Center

Daily	Weekly	Monthly
pH of System	Bacteria Population	Corrosion Coupon Checks
Turbidity	Corrosometer Readings	Electrical Consumption
Conductivity	Water Temperature	Maintenance & Monitoring Costs*
Ozone Concentration		
Visual Inspection		

*Were accumulated as expended and summarized monthly

Data Source - Kennedy Space Center

b.   Ozone is effective in controlling microbial growths in cooling tower water systems at concentrations of 0.05 to 0.20 ppm in the basin.

c.   Ozone can inhibit or prevent scale deposit in the heat exchanger tubing, water boxes, cooling tower fill and structure.

d.   Corrosion rates for steel range between 3 to 5 mils/year.

e.   Ozone meets or exceeds the effectiveness of the standard chemical methods.

f.   Operating and maintenance using ozone are about 30 percent of the costs of using chemicals.

Field Experience Comments

There are many problems that can and do occur in the field. Such problems are not necessarily covered by the manufacturer's instructions; therefore, the following suggestions might be useful:

1.   The make-up water should be free from noticeable sediment, mud, discoloration, etc.

2.   The material in the ozone treated system should be compatible with ozone. The ozone line from the ozone generator to the gas/water contactor carries the highest ozone concentration (1 to 4 percent); therefore, the line material should be stainless steel or CPVC.

3.   The ozone generator must be checked for ozone leaks during operation. Such leaks could damage the controls, component and ozone generator cabinet.

4.   For efficient operation, the ozone generator should be located in an air-conditioned area. Excessive heat (90°F–32.2°C) might damage the system or reduce the ozone generating capacity.

5. PVC is ideal for the main water recirculation system, from the tower through the heat exchanger and back to the tower.

6. During the first 4 to 6 weeks of the ozone treatment, the sump of the cooling tower will accumulate scale that has been removed from the fill, louvres, tower structure, condenser/heat exchanger tubing and water boxes. This material should be removed either by vacuum or shovel.

7. The actual capacity of the ozone generator should be certified by the manufacturer and checked yearly.

8. Corrosion coupons for both copper and steel should be placed in the system and checked at least every 6 months. This needs to be done in every system regardless of the water treatment method.

9. Certain parameters of operation should be checked on a routine basis. These parameters are shown in Figure 9-2 and Table 9-4.

10. The compressor power meter readings should be taken before and after the ozone treatment is started for determining the actual annual power saving.

U.S. Ozonair Corporation furnished ozone equipment for most of the cooling towers that have been ozonated. Mr. Karel Stopka, president of U.S. Ozonair comments as follows:

1. Ozone equipment furnished to 84 cooling tower systems, some of which have operated since installation without any blowdown, for as long as 4 years.

2. Some of the cooling tower systems using ozone water treatment that have gone for as long as 4 years without blowdown are:

   a. City Hall, Hayward, CA, installed in 1982—6,000 total water capacity—32 grams/hour ozone requirement.

    b. Lawrence-Livermore Laboratory, University of California, installed in 1983—5,000 gallons total water capacity—32 grams/hour ozone was required.

    c. GTE, Mountain View, CA, installed in 1982.

    d. Akron Porcelain and Plastics, Akron, Ohio—installed in 1984—32 grams/hour ozone required.

3.   IBM, T. J. Watson Research Center, Yorktown, NY installed an ozone treatment system in June 1985 in order to make corrosion and scale studies. Six different corrosion coupons were installed in the lines. On December 6, 1985, these coupons were removed from the line for inspection. All of the coupons were found to be clean and free of corrosion and pitting. IBM has requested funding for two large cooling tower systems for the installation of ozone water treatment systems. Currently, IBM has a cooling tower water treatment system which is undergoing further testing with ozone.

4.   After the excellent results of the testing at the KSC, funds have been requested for the installation of ozone water treatment systems for all of their cooling towers. KSC has purchased additional ozone equipment for their cooling towers.

Stopka offers the following suggestions regarding the installation of ozone equipment in cooling tower systems:

1.   When calculating the ozone requirements for a cooling tower system, you must include treatment of all of the water in the system and not just the recirculated water. In a system recirculating 1,000 gpm with a system total of 5,000 gallons, the ozone equipment must be sized for the treatment of the 5,000 gallons, not only the 1,000 gallons.

2.   The make-up water should be of a high quality, not wastewater.

3. The dew-point of the air entering the ozone generator must be $-50°F$ ($45.6°C$) or lower, preferably $-60°F$ ($51.1°C$). No moisture must ever enter the ozone generator.

4. The ozone generator should be sized for the heaviest load demand. Summer demands are more severe than winter demands.

5. The quality of the air entering the cooling tower influences the performance of the water treatment system. Units located above ground level should have better air quality than those located at ground level.

Conclusions

Ozone should be a viable alternative to the use of multiple-chemicals for the treatment of cooling tower water, because:

1. No bleed-off (blow-down), thus reducing the water costs and sewage charges.

2. No pH control.

3. No chemical costs, air is free.

4. Ozone is an excellent biocide—reducing or eliminating bacteria, including the Legionnaires' Disease Bacterium; bioslime; algae; mildew; and mold.

5. Corrosion rates are 50 percent lower for both copper and steel when using ozone.

6. Air and water effluent with EPA regulations.

7. Ozone, as the second strongest (next to fluorine) oxidizing agent, will oxidize organic and some inorganic material.

8. Ozone, by the elimination of biomass, can reduce or eliminate scaling in the condenser tubing, on the louvres and fill

and scale on the cooling tower structure. By reducing the scale formation in the condenser/heat exchanger tubing, the fouling factor can be reduced. If the design fouling factor is 0.002 but decreased to 0.0005 by using ozone, the condensing temperature would be reduced from 107 to 99°F (41.7 to 37.3°C) with a possible savings of 16 percent in compressor horsepower and a 6 percent increase in compressor capacity. The process heat exchanger could have a 200 percent increase in effective surface.

9. Inorganic materials, such as iron, aluminum and manganese are oxidized to hydroxide forms that can act as coagulants.

10. Savings of 50 percent or more in operating and maintenance costs.

11. Since ozone is generated at the job site from air, there is no hazardous material storage.

12. An ozone line breakage could be hazardous.

13. Due to the biological potency of ozone, bacterial delignification in wooden towers should be materially reduced. Also, because there is no chlorolignin nor a combination of alkaline, pH and chlorine in the water would help deter delignification.

14. By reducing or eliminating the bio-slime layers which provide favorable anerobic conditions for sulfate-reducing bacteria, the potential corrosion rate would be reduced.

15. Water will be conserved, energy to the refrigeration compressor will be reduced, chemical costs will be greatly reduced or eliminated and air and water effluent will comply with EPA regulations.

16. Savings in owning and operation costs will cover the capital outlay for the ozone system in from 6 to 18 months, depending upon the size and complexity of the system.

Recommendations

In view of the information and data presented here, the following suggestions are offered:

1. Although ozone has been proven to be an excellent alternative for the standard method of treating cooling tower water, there has not been sufficient documentation to establish any criteria.

2. Recommended parameters for ozone testing could be used to test wood samples for determination of ozone effectiveness in the prevention of delignification.

3. The Cooling Tower Institute (CTI) should recommend that all ozone generators and ancillary items be certified for capacity by the manufacturer. All ozone generators should be given an approved capacity test prior to shipment (if the capacity is 10 grams/hour or larger). In addition, the method of capacity testing should be stated and the certificate of capacity and equipment warranty should accompany each unit.

Legionnaires' Disease Bacterium (LDB)

During the 1976 American Legion Convention in Philadelphia, 221 cases of a severe pneumonia occurred with 34 fatalities. The disease was identified as a bacterial-type of pneumonia, caused by Legionella pneumophila. Since that time, many cases of Legionellosis have been identified. Epidemiological evidence has implicated cooling towers and evaporative condensers as the common source of infectious Legionella in some outbreaks of Legionellosis. This disease has been identified in many localities in the U.S. and throughout the world. Estimates of the number of cases of pneumonia attributed to Legionella infection range from 25,000 to 200,000 per year. Currently there are many fatalities each month in the U.S. from this disease. The case fatality rate can reach 15 to 45 percent in untreated or improperly treated individuals.

Ozone as a Disinfection Agent (Biocide)

Ozone has been cited as displaying an "all or nothing" kill effect on bacteria. The high oxidation potential of ozone will explain this effect. This effect can be attributed to the high oxidation potential of ozone. Ozone is such a strong biocide that only a few milligrams per liter are required to measure germicidal action. Water that contains bacteria may contain dissolved organic and/or inorganic material which can react with ozone, creating an initial ozone demand, thus delaying biocidal action. The kill rate will be very fast after this initial stage. These observations are the cause of the "all or nothing" effect.

Table 9-5 shows the comparative disinfection efficiency of ozone relative to various chlorine species. This data clearly shows that ozone is a more powerful biocidal agent against organisms by a factor of 5 to 100.

Table 9-5. Comparative Disinfection Efficiency

Disinfectant	Enteril Bacteris	Virus	Spores	Amoebic Cycts
Ozone (O_3)	500	0.5	2	0.5
Cl_2 as HoCl	20	1.0	0.05	0.05
Cl_2 as OCl	0.2	0.02	0.0005	0.0005
Cl_2 as Hn_2Cl	0.1	0.005	0.02	0.02

Where I = (Mg/l) (Min) at 5°C (41°F)
= Specific susceptibility coefficient when organisms are compared
= Specific lethality coefficient when germicides are compared

Data Source - PCI

Table 9-6 shows the extraordinary low dosage of ozone required to inactivate viruses and bacteria in ozone demand-free water. When bacteria or viruses are associated with inorganic materials, such as alum floc or bentonite clay, the ozone dosage level required for disinfection is independent of their being free or associated. Alum floc and bentonite clay have no ozone demand whereas organic substances exert a substantial ozone demand. The complete destruction of virus after they have become associated with a bacteria cell may require ozone dosage as high as 5 mg/l (ppm). Ozone at sufficiently high concentration, not only disinfects, it sterilizes (destroys) almost all forms of life, including spores and cyctes. Data indicates that ozone should be the last step in the water treatment because there are lower organic levels of concentration to compete with ozone during the disinfection process. Surface water, usually, requires ozone residual in the 0.1 to 0.5 mg/l range with contact time between 1 to 6 minutes. Applied ozone dosage ranges from 1.5 to 2.0 mg/l and can increase to 11 mg/l for some surface water.

Ozone Generation

Ozone is generated from ambient air or oxygen. The air is dried to a dew point of 60°F (−51.1°C) or lower by an electric apparatus at the job site. Oxygen used as the carrier gas, will produce about twice as much ozone as is produced from air using the same machine and power. The concentration of ozone generated from air ranges from 1 to 2 percent concentration, while the ozone generated from oxygen will range from 2 to 4 percent concentration. The concentration of ozone that is generally used in cooling tower application is generated from air at between 1 and 1.5 percent (10,000 to 15,000 ppm). Ozone generating capacity is given as either grams of ozone/hour or as pounds/day. 18.9 grams/hour is equal to 1 pound/day.

Figure 9-1 shows the location of the various components of a cooling tower ozonation system. Figure 9-2 indicates the suggested controls or monitors used in the ozonation system of the cooling

Table 9-6. Disinfection of Microorganisms in Ozone Demand Free Water at 20°C (69°F) and pH 7

| Microorganisms | Ozone Applied (Mg/l) | Ozone Residual After 40 Seconds | Initial Virus Concentration (Pfu/Ml) | Percent Inactivation at Given Contact Time In Seconds | | |
				10	20	30
Porcine Pigorna Virus	0.024	0.017	1.1×10^1	100	100	100
F_2 Bacteriophage	0.33	0.28	1.3×10^7	99.99+	99.99+	99.99+
Coxsasckie Virus	0.51	0.023	1.5×10^4	99.98	100	100
Polio Virus	0.012	0.004	5.1×10^2	100	100	100
Polio Virus	0.015	0.011	1.4×10^4	99.6	99.98	99.98
Escherichia Coli	0.53	0.49	2.0×10^8	99.99+	100	100
Escherichia Coli	0.239	0.174	9.8×10^7	99.999	99.999	100
Escherichia Coli	0.57	– –	3.6×10^{10}	99.992	99.999	100

Data Source - PCI

tower. These controls/monitors are recommended by the NASA/Jet Propulsion Laboratory tests.

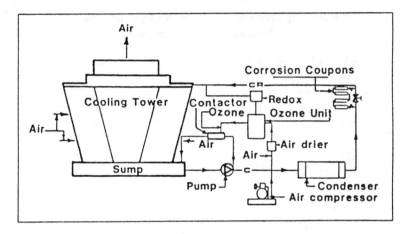

Figure 9-1

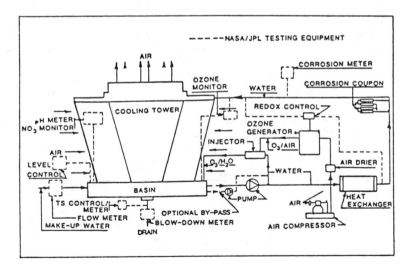

Figure 9-2. Cooling Tower Ozonation System and NASA/JPL Testing Equipment

Table 9-7 shows the operating and maintenance costs associated with the Kennedy Space Center testing programs.

Table 9-7. Operating and Maintenance Costs per Year (KSC)
Chemical vs. Ozone Treatment

Item	Chemical	Ozone System
I. CHEMICALS:		
Corrosion Inhibitor	$1800.00	0
Microbiocide A	$ 810.00	0
Microbiocide B	$ 250.00	0
II. WATER CONSUMPTION		
Blowdown, gallons	1,260,000	0
dollars	$1260.00	0
III. ELECTRICAL COST:		
kWh (dollars)	0 (0)	5280 ($265.00)
IV. LABOR: (See Note)		
*Maintenance, hrs ($)	80 ($1680.00)	32 ($675.00)
*Field Monitoring, hrs ($)	130 ($2860.00)	50 ($1050.00)
*Lab Monitoring, hrs ($)	25 ($525.00)	75 ($1575.00)
Chemical Additions, hrs ($)	130 ($2860.00)	0
Yearly Total, ($)	$12,045.00	$ 3,565.00
Cost of Ozone System (Installed)		$16,057.00

*These costs apply in the case of testing purposes.
Electricity, $265/year (only operating cost).

Data Source - Kennedy Space Center

A final precaution is to be noted: with respect to treatment of cooling tower waters, temperatures should be limited to 130°F, or even to 120°F. While the reasons for this limitation are not quite clear at this time, above these maximums, the ozone does not perform to the anticipated extent.

ULTRAVIOLET

In conjunction with activated carbon filters, and demineralizers, a problem may be experienced with biological growths in the effluents. As these effluents are usually pure, chemical disinfection may not be the ideal answer. Many such conditions may be treated by the use of Ultraviolet (UV) radiation. UV disinfection has become a viable alternative to chlorine and ozone. UV is nonchemical, and does not form toxic by-products like trihalomethanes (THMs). It is effective on both pathogenic bacteria and viruses with relatively smaller dose latitudes. It is not pH dependent, and does not produce toxic disinfectant residuals.

Simple and efficient as UV is many users, however, have been disappointed with operating results. These poor results may generally be traced to:

- Insufficient dosage
- Inefficient UV spectrum selection
- Not enough power
- Poor hydraulic performance
- Excessive absorbance
- Unreliable lamps

UV lamps are similar to fluorescent-tube lamps except that the tube is made of quartz glass, and has no phospher coating inside the tube. Basic lamp enclosures are annular and coaxial. The annular design has the UV lamp enclosed in a quartz glass sleeve immersed in water, and the coaxial design has the water flowing through quartz glass or Teflon tubes, irradiated by external UV lamps. The spacing of the lamps should be designed to insure that 90 percent of the useful UV energy from each lamp is used in multiple lamp reactors. Spacing should not be too close, since the lamps will absorb most of the UV energy and will overheat. The lamps, filled with low pressure mercury vapor and argon, emit rays with wavelengths between 200 and 290 nanometers (nm), with maximum effectiveness around 260 nm, depending on the organism. Figure 9-3 is an illustration of the UV spectrum of interest.

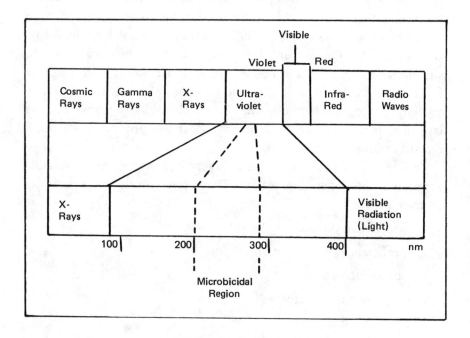

Figure 9-3. UV Spectrum
Source: Aquionics, Erlanger, KY

The available germicidal output of a UV lamp at 260 nm from a low pressure mercury vapor lamp is about 20–25 percent of the lamp rating. Manufacturers should be consulted and rated average UV output after about 100 hours of operation should be used as the basis for dose density calculations.

Lamp switching should be minimal to reduce the solarization or darkening of the glass tubes, and to minimize a rise in temperature. Medium and high pressure UV lamps can withstand higher temperatures, but their output is less than for low pressure lamps.

Lamp intensity and dosage is calculated in watts, depending on the size and power of the lamps. The manufacturer should state how much germicidal UV is emitted at 260 nm. UV dosing is calculated as the product of the exposure intensity and the exposure time. Thus:

$$\text{Germicidal UV Dose} = 1\,(\mu W/cm^2 \text{ at } 253.7 \text{ nm}) \times T_e(\text{secs})$$
$$(\mu W\text{-sec}/cm^2)$$

For organisms found in secondary wastewater effluent prior to disinfection, 16,000 μW-sec/cm$\oplus$ is sufficient to inactivate 99.9+% of microorganisms.

The exposure time should be based on the intensity at 90 percent absorbance. UV in water is absorbed by dissolved organisms in the water, and suspended solids. The degree of absorbance depends of course on the coloration and turbidity of the water. If water is not entirely pure, as, for example, the effluent from DI systems, filters should be installed ahead of the lamps. Absorbance may be calculated as follows:

$$I_1 = I_0 10^{-ad}$$

where: I_1 = measured intensity at distance d in cm
I_0 = lamp surface intensity
a = medium abosrbance (cm^{-1})

Effluent quality of 30:30 or more may be acceptable if absorbance does not exceed 0.2 cm^{-1}. A review of secondary effluents throughout the U.S. indicates that absorbance did not exceed 0.2 cm^{-1}. This absorbance value of 0.2 cm^{-1} can be used to size UV reactors based on the flow layer absorbing 90 percent of the germicidal energy.

In calculating or checking performance, a safety factor of 2 should be provided. Flow should be sufficiently high to reduce any scale deposition. Depending on the TDS of the water, velocity should be 1 fps or greater. Theoretical exposure time is usually substantially less than theoretical retention time.

The hydraulics of UF reactors is important in applying dosage. Plug flow (zero dispersion) should be the goal in designing the flow system, since ideal plug flow has minimum dispersion. The lower the dispersion factors (expressed in cm^2/sec), the better the UV efficiency. In the design, the Reynolds Number, R_e, should be above 2000 in the average flow channel to assure turbulent flow. Turbulent flow with minimum dispersion will apply a minimum effective germicidal dose to all organisms in the flow path. Small dispersion assures that any group of organisms in the reactor will get the design dose, plus or minus ten percent. Little energy is wasted if some organisms get 110 percent of dose. With greater dispersion the system approaches mixed flow conditions, with the result that any group of organisms may receive design dose plus or minus 50 percent. When the vairation in dose rate is 50 percent, the lamp intensity at 50 percent should be the minimum design dose. Various flow parameters should be specified to ensure that the installation has plug flow characteristics. These include R_e above 2000, the ratio of length to hydraulic radius R_h, over 50, uniform velocity, uniform cross sectional velocity, and minimum amount of air bubbles at open inlets.

There are no microorganisms known to be resistant to UV which, unlike chlorination, is highly effective against bacteria, viruses, molds and yeasts. In practice, bacteria and viruses are the cause of the major waterborne pathogenic diseases. Of these, enteric viruses, hepatitis virus and Legionnella pneumophila have been shown to survive for considerable periods in the presence of chlorine, but are readily eliminated by UV treatment. Similarly, chlorine forms trihalomethanes in surface water whereas UV treatment does not produce toxic by-products.

Concentration-Residence Time Distribution (RTD) curves will indicate the dispersion characteristics of the system. An experimental method uses a salt tracer input pulse to generate an RTD curve. Manufacturers should be required, on larger systems, to provide such a curve for each flow condition.

Table 9-8 details the advantages or disadvantages of various disinfection systems.

**Table 9-8. Advantages of UV Systems
Disinfection Methods Comparison**

	Ultra Violet	Simple Chlorination	Chlorination/ Dechlorination	Ozone
Capital cost	Low	Lowest	Medium	High
Operating cost	Lowest	Low	Low	High
Ease of installation	Excellent	Good	Complex	Complex
Ease of maintenance	Excellent	Good	Good	Poor
Cost of maintenance	Low	Low	Medium	High
Frequency of maintenance	V. infrequent	Frequent	Frequent	Continuous
Disinfection performance	Excellent	Some regrowth possible	Some regrowth possible	Unreliable in effluent
Virucidal effect	Good	Poor	Poor	Good
Personnel hazards	Low to none	High	High	High
Toxic chemicals	No	Yes	Yes	Yes
Effect on water	None	Forms trihalomethanes	Forms trihalomethanes	Toxic by-products
Residual effect	No	Yes	Yes	Some
Problems with Operating Systems	Low	Medium	Medium	High
Contact time	1-5 secs	30-60 mins	30-60 mins	10-20 mins
Ease of handling varying flow rate	Excellent	Poor	Poor	Good

Addressing the question of reliability, the operators of the system must

- Set up to maintain desired flow without excessive switching
- Maintain water in the tubes, or over them, at all times
- Monitor the lamps continually
- Clean tubes on a regular basis
- Provide intensity meters and lamp failure indicators
- Provide alarms for system failure, low UV, overtemperature
- Install a chamber temperature cut-out thermostat.

10

Special Considerations in Boiler and Cooling Water Plants

PROTECTION OF CONDENSATE PIPING

Condensate lines in power plants are frequently subject to heavy corrosion for a number of reasons:

- Low pressure may allow the entrance of air via pump seals and fittings

- Excessive alkalinity in the boiler system causes the release of Carbon Dioxide (CO_2) when the steam is condensed to a lower pressure

- The elevated temperatures in the condensate system accelerate the action of oxygen or CO_2.

Control of oxygen corrosion is accomplished by eliminating the air, or by the use of filming amines which provide some protection of the pipe material.

Control of carbon dioxide can be achieved in various ways. Some of the methods include the following:

1. *Removal of free carbon dioxide from boiler feedwater and/or elimination of carbonate and bicarbonate alkalinity.* Free carbon dioxide can be removed from boiler feedwater by de-aeration prior to entering the boiler. However, carbonate and bicarbonate alkalinity do not decompose to the corrosive gas until it reaches steam drum temperatures. Methods such as anion demineralization, chloride dealkalization, hydrogen zeolite softening followed by a decarbonator, or cold/hot lime softening are utilized in reducing the level of alkalinity in the feedwater. Even with good dealkalization of the boiler feedwater, some CO_2 will enter into the steam. Low levels of the gas can be corrosive and should be chemically controlled.

2. *Use of volatile neutralizing amines that react with the acidic ions to elevate the condensate pH.* For example, Cyclohexamine or Morpholine. These amines react chemically to neutralize the carbonic acid—thus raising the condensate pH to a value that will ensure low metal corrosion rates. The feed requirement of a neutralizing amine is directly proportional to the amount of carbon dioxide present in a system. The amine is usually fed to maintain a condensate pH in the range of 8.0-9.0.

The selection of a neutralizing amine is dependent on a number of factors—among these are the following:

a. The amine stability must be such that the amine will not decompose under normal operating temperatures and pressures.

b. The basicity of an amine is a measure of both the relative amount of acid that can be neutralized by the amine and the highest pH attainable with that amine. Therefore, an amine with a high basicity is desirable.

c. When the amine reacts with the carbonic acid in the condensate, an amine bicarbonate is formed. This bicarbonate must be soluble in condensate to prevent deposition in piping and equipment.

d. The amine must also have sufficient volatility so that it will be present in the steam and, therefore, be able to react with the dissolved carbon dioxide in the condensate. The volatility is characterized by the distribution ratio (DR).

The DR is an important characteristic in choosing an amine for industrial usage. It is defined as the ratio of amine in the steam to that amount found in the bulk fluid or condensate. Since protection is dependent upon condensing the amine at various system temperatures and pressures, the DR and consequently the amine must be selected with operating conditions taken into consideration.

To provide good protection against corrosion at all points in a condensate system, a corrosion inhibitor must be properly distributed in the system. An amine with a low DR, such as morpholine, will tend to be present in greater concentration in the condensate than in the steam. Therefore, this amine would be considered in a high pressure operation where turbine protection is essential. Many plants have low pressure systems with extensive condensate return lines. In this instance, an amine with a high distribution ratio such as cyclohexylamine would be preferred. The high DR indicates that the amine will tend to remain in the steam and be carried to the far end of the condensate system where high concentrations of carbon dioxide accumulate. Other amines and amine blends commonly used in this application have DRs which fall between those of morpholine and cyclohexylamine and are applied subject to plant requirements.

Ammonia was one of the first agents used to control corrosion by pH neutralization and elevation. However, as a result of the high distribution ratio at deaerator conditions, significant amounts of ammonia would be lost through the vent. Ammonia in high concentrations attacks copper and copper alloys, particularly in the presence of oxygen. For these reasons, special expertise is needed in the application of ammonia as a corrosion control agent.

In many modern industrial facilities, the steam condensate system can be quite complex. They may include equipment and heating systems operating at different pressures. These plants require the protection of an amine with a low DR for initial condensation and a neutralizing amine with a higher DR to protect the extensive run of steam lines. Under these circumstances, the best protection is provided by blending amine products containing a variety of materials with different distribution ratios.

Blended amines should be chosen with consideration to the operating parameters and system conditions. The optimum distribution ratio required by a plant is a function of the process steam pressure and the percent condensate formed at the turbine.

When this ratio is determined, a neutralizing amine blend can be selected based on the blend characteristics. These characteristics include a number of factors—the most important being the pressure dependent distribution ratio. By selecting a neutralizing amine on a technical basis, total system protection can be obtained.

Many advantages are afforded by use of neutralizing amines. Neutralizing amines are easily controlled since their feedrate is based upon pH analysis of the condensate. Water dilutions of amine products are compatible with most internal feedwater conditioning chemicals. They can be fed along with internal treatment, requiring no special feed equipment or feed point. In some cases, particularly where treatment is only needed in a portion of the system, amine may be fed directly to the steam line using proper feed equipment. Since pure amines have low flash points, the materials are normally supplied in aqueous solutions which are quite safe.

Amine application to neutralize dissolved carbon dioxide will not offer direct protection against oxygen attack. However, extensive damage can be averted if low oxygen levels are maintained and the pH of the condensate is above 8.0.

3. *Application of filming amines which distribute a protective coating on metal surfaces, protecting the piping from carbon dioxide attack.* For example, Octadecyclamine or Hexadecyclamine. These high molecular weight organic materials are branched- or straight-chain amines designed to form a protective film on condensate metal surfaces. Filming amines

were developed as an economical means of controlling attack by carbon dioxide. The advantage of filmers is that the treatment feedrate does not depend upon the amount of corrosive gas in the system. The protective film is also effective in controlling oxygen corrosion when this gas is present in the condensate system in small quantities.

Filming amines function by forming a monomolecular film on the base metal of the condensate system. One end of the highly polar amine molecule attaches itself to the metal surface while the other end provides the hydrophobic, or water repelling, barrier between the system metal and the aggressive condensate. Effective filming amines form a closely packed amine monolayer that is adsorbed onto the surface at low amine concentrations. Branched chain compounds are less efficient inhibitors than are straight chains because the bulky groups on branched chains prevent tight film formation. The adsorption properties of filming amines must also be considered. Increasing the chain length of a primary amine (straight chain) increases its adsorption characteristics.

The use of filming amines has some benefits over neutralizing amines. Except in systems where carbon dioxide is exceptionally low, filming amines are more economical. The dosage rate of filmers is dependent upon the amount of steam produced and, to a lesser degree, the expense of the condensate system. This fixed concentration of amine develops and maintains the desired protective film. Another advantage afforded by filming amines is their ability to effectively protect the metal surface against attack by oxygen. Due to the formation of a film barrier, dissolved oxygen diffusion to the iron or copper surface and, consequently, metal oxide formation is retarded.

Determination of the optimum feedrate of a filming amine is sometimes difficult. A high feedrate can induce thick, sticky amine deposits. This overfed amine may also return to the boiler where the amine can add to the sludge load. However, low feedrates of amine will cause ineffective film formation and corrosion will occur.

In many instances where the system is extensive, the problem is solved by supplemental feed points.

Filming amine deposition can be caused by contamination of the condensate. Severe contamination will cause the concentration of effective filming amine to fall below the amount required to repair or replace a film. Oil, hardness and oxidizing substances cause the most serious fouling. When oil is allowed to enter the condensate system, it acts as a solubilizing agent and reacts with the amine stripping away the protective film barrier. Sulfates and chloride associated with hardness contamination react with the filmer to form sticky deposits. In other cases, oxidizing agents cause polymerization of amines and reduce film effectiveness. Therefore, it is imperative to keep the condensate systems free of contamination since this can hinder the performance of filming amines in a return line system.

Cost considerations, as well as technical aspects of an amine program, will be important in selection of the treatment method. In some applications just a neutralizing or filming amine approach does not provide sufficient protection. This can be due to higher than normal levels of oxygen intrusion or the complexity of the plant where filming amines do not reach all areas of the condensate system.

One effective approach has been to combine primary filming amines with neutralizing amines. With these treatment programs, the neutralizing amine can be fed to eliminate the pH and/or to carry the filming amine to remote areas of a complex condensate system.

When starting amine treatment in an old condensate system, care should be exercised in the intensity of the treatment. To prevent sloughing of badly fouled condensate lines and return of corrosion products to the boiler, the amine should be initially fed at a low feedrate and gradually increased to recommended treatment levels. During this period, soluble and total iron levels in the condensate should be monitored so that control of the system cleanup is maintained.

OXYGEN DAMAGE

Where pitting and thinning are apparent in condensate piping, oxygen attack is present.

Oxygen does not generally present a major problem as a corroding agent in the condensate. However, small leakages from receivers, vacuum condensers, and other sources may occur and result in oxygen corrosion. This type of attack is easily recognized by a broad, pitted region with a red oxide layer surrounding the pit. The presence of black iron oxide along the concave area of attack indicates an active corrosion site. In most cases, where small amounts of oxygen enter the system, extensive damage will not occur. However, oxygen even in low concentrations worsens carbon dioxide attack by initiating the corrosion cycle. Oxygen can start the cycle by creating a pit. The low pH condensate resulting from solubilized carbon dioxide will attack the bare surface and remove the protective magnetite film surrounding it. The metal is then exposed to additional oxygen and the cycle is repeated.

REVIEW OF FEEDWATER
QUALITY LIMITS

In earlier chapters we have discussed quality limits established for boiler water. The tabulations following represent the latest thinking with regard to reliable operation of boiler systems. For convenience, several tabulations of limitations are included here, for comparison of one with another. Figures 10-1 through 10-3 and Table 10-1 are from a 1989 technical paper written by J. Lux of the Babcock & Wilcox Company. Tables 10-2 through 10-4 are from recent ASME publications.

Desalination Systems, Inc. of Escondido, CA offer specialized information in Table 10-5. To illustrate the need for silica removal, the table lists maximum silica and total dissolved solids (TDS) limits for the power and electronics industries. As shown, limits set for silica are much more stringent than those established for TDS.

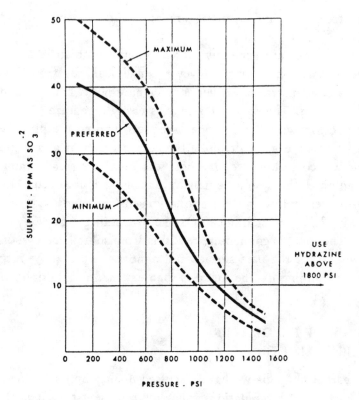

BOILER WATER LIMITS

Figure 10-1
Sulphite vs Boiler Pressure

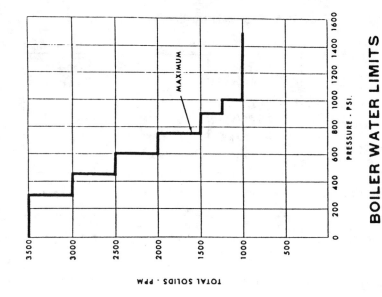

BOILER WATER LIMITS

Figure 10-3
TDS vs Boiler Pressure

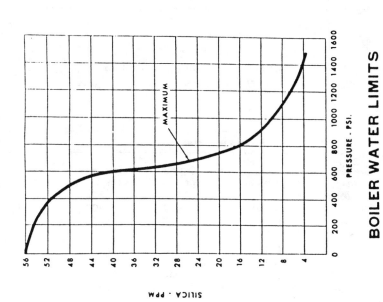

BOILER WATER LIMITS

Figure 10-2
Silica vs Boiler Pressure

Table 10-1. Feedwater Quality Limits

	BELOW 600 psi	600 TO 1000 psi	1000 psi & ABOVE
TOTAL HARDNESS (AS PPM CaCO3)	0 (1 PPM TEMPORARY MAX.)	0 (1 PPM TEMP. MAX.)	0
IRON	0.1 PPM	0.05 PPM	0.01 PPM
COPPER	0.05 PPM	0.03 PPM	0.005 PPM
OXYGEN	0.007 PPM	0.007 PPM	0.007 PPM
pH	7.0 TO 9.5	8.0 TO 9.5	8.5 TO 9.5

Table 10-2. Water-Quality Guidelines Recommended for Reliable,
Continuous Operation of Modern Industrial Watertube Boilers

Boiler feedwater			
Drum pressure, psig	Iron, ppm Fe	Copper, ppm Cu	Total hardness, ppm CaCO3
0-300	0.100	0.050	0.300
301-450	0.050	0.025	0.300
451-600	0.030	0.020	0.200
601-750	0.025	0.020	0.200
751-900	0.020	0.015	0.100
901-1000	0.020	0.015	0.050
1001-1500	0.010	0.010	ND[4]
1501-2000	0.010	0.010	ND[4]

Table 10-3. Maximum Limits for Boiler Water

Boiler Pressure (psig)	Total Solids (ppm)	Alkalinity (ppm)	Suspended Solids (ppm)	Silica* (ppm)
0-300	3500	700	300	125
301-450	3000	600	250	90
451-600	2500	500	150	50
601-750	2000	400	100	35
751-900	1500	300	60	20
901-1000	1250	250	40	8
1001-1500	1000	200	20	2.5
1501-2000	750	150	10	1.0
Over 2000	500	100	5	0.5

*Silica limits based on limiting silica in steam to 0.02-0.03 ppm.

Table 10-4

Boiler water

Drum pressure, psig	Silica, ppm SiO_2	Total alkalinity,[1] ppm $CaCO_3$	Specific conductance, μmho/cm
0-300	150	350[2]	3500
301-450	90	300[2]	3000
451-600	40	250[2]	2500
601-750	30	200[2]	2000
751-900	20	150[2]	1500
901-1000	8	100[2]	1000
1001-1500	2	NS[3]	150
1501-2000	1	NS[3]	100

[1]Minimum level of hydroxide alkalinity in boilers below 1000 psi must be individually specified with regard to silica solubility and other components of internal treatment.

[2]Maximum total alkalinity consistent with acceptable steam purity. If necessary, the limitation on total alkalinity should override conductance as the control parameter. If make-up is demineralized water at 600-1000 psig, boiler water alkalinity and conductance should be as shown in the table for the 1001-1500 psig range.

[3]NS (Not specified) in these cases refers to free sodium- or potassium-hydroxide alkalinity. Some small variable amount of total alkalinity will be present and measurable with the assumed congruent control or volatile treatment employed at these high pressure ranges.

Table 10-5

Industry	Maximum Contaminant Level, Mg/l SiO_2	TDS
Electronics (Rinse water)	0.010	0.0277
Power Industry (Boiler water)		
Drum pressure, psi		
0 - 300	150	3500
301 - 450	90	3000
451 - 600	40	2500
601 - 750	30	2000
751 - 900	20	1500
901 - 1000	8	1250

The above boiler water values reflect ABMA and ASME water quality guidelines for boiler blowdown quality. Energy savings can be achieved by minimizing the feedwater levels of both SiO_2 and TDS. This results in less scale formation, lower chemical consumption and reduced boiler water blowdown.

CORROSION AND DEPOSITS IN COOLING SYSTEMS–THEIR CAUSES AND PREVENTION[1]

Cooling water problems are becoming more and more associated with the greater problems that are faced during the planning and operation of plant systems. The ever-increasing industrialization is primarily responsible for the gradual deterioration in the quality of

---

[1] Abstracted from The Sulzer Technical Review, March 1972. Published in *Combustion* Magazine, 1973.

water, particularly surface water. Industry is directly affected by the state of affairs, for it is continually compelled to use water whose natural properties deviate perpetually from the ideal condition required for cooling purposes. The difficulties that consequently arise in the operation may be classified fundamentally into two different groups:

- stimulated corrosive effect on metallic constructional materials and/or
- scaling and fouling, particularly in pipe systems and on heat exchanger surfaces, which consequently impairs the system efficiency.

In principle, there are three possibilities of counteracting this danger, namely:

- selection of a "good" cooling water. Admittedly, the selection of same is already predetermined by the geographical location of the plant in most cases.
- selection of more suitable (and mostly more expensive) materials.
- suitable water treatment.

Before discussing the specific problems of cooling systems, it would appear to be more expedient to express the term "water quality" more explicitly to begin with, and also provide some information regarding the considerations that have to be made when classifying cooling water.

The Term "Water Quality"

Cooling waters are never pure water, H_2O. On the contrary, they are nearly always diluted aqueous solutions of very different composition. The waters or coolants can be subdivided into:

Natural Waters
  - ground and surface water
  - sea and brackish water

Treated Waters
  - softened or deionized water

Special Cooling Liquids
  - cooling brines

Water is classified by means of chemical analysis, the purpose of which is to determine the presence of certain dissolved individual substances, i.e. the cation content (Ca^{++}, Mg^{++}, Na^+, K^+, Fe^{++}/Fe^{+++}, Mn^{++}, NH_4^+, H^+), the anion content (HCO_3^-, CO_3^{--}, SO_4^{--}, Cl^-, NO_3^-, PO_4^{3-}, OH^-) as well as the dissolved oxygen content.

Natural Waters

Determinative factors for the classification of a water are:

pH Value

The pH value, which is a measure for expressing the hydrogen ion concentration, is one of the most important measuring characteristics of a water. Natural waters produce more or less neutral reactions (pH value about 7); soft waters, which contain a great deal of free carbon dioxide, and moor waters, which have a low salt content and a great deal of carbonic and humic acid, tend to have a slight acidulous effect (pH value between 5 and 6). Waters with even stronger acidulous reactions (pH value <5) are very rarely found in the natural state. Generally speaking, we can say that neutral and weak alkaline waters usually cause no or only very slight corrosion on metallic materials, but one can nearly always expect corrosion in weak to strong acidulous areas.

Nevertheless, the pH value is not the only determining factor, for the water hardness—particularly the carbonate hardness—is of decided importance.

Water Hardness

The hardness of a water is determined by its calcium and magnesium salt content. We differentiate between three different kinds of water hardness:

- *Total Hardness* (TH), i.e. the total of hardness-forming calcium and magnesium ions.

- *Carbonate Hardness* (CH), i.e. the calcium and magnesium ions that are combined with the bicarbonate ion HCO_3^-.

- *Noncarbonate Hardness* (NCH), i.e. the percentage of calcium and magnesium ions which are allocated to an anion other than the bicarbonate. In most cases, these are mostly alkaline earth sulphates or chlorides.

The hardness is expressed in degrees of hardness, whereby one degree of the German hardness scale ($1°$ dH) corresponds to 10 mg CaO/l and one degree of the French scale ($1°$ fH) represents 10 mg $CaCO_3$/l. A natural water always contains free carbon dioxide, in larger or smaller amounts, together with HCO_3^- ions, formed by the dissociation of carbonic acid (Figure 10-4).

We define the associated free carbon dioxide which is required to retain the alkaline earth carbonate in solution in accordance with the equation

$$CaCO_3 + CO_2 + H_2O \rightleftarrows Ca^{++} + 2HCO_3^- \qquad (1)$$

The associated free carbon dioxide can be ascertained from one of Tillman's[1] experimentally determined equations, according to which

$$CO_2 \text{ (associated, free)} = (CO_2 \text{ combined})^2 \cdot CaO \cdot K \text{ (mg/l)} \qquad (2)$$

whereby $\log K = 0.01259 \cdot t - 4.92670$ (t = temperature $°C$)

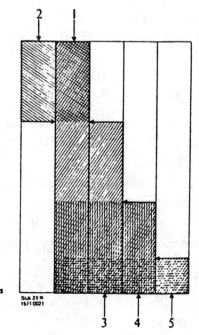

1 Total CO_2
2 CO_2 as bicarbonate
3 Free CO_2
4 Rust-promoting CO_2
5 CO_2 agressive to calcium carbonates

Figure 10-4. Schematic Subdivision of the Total CO_2 into Free, Connected, Associated and Rust-Promoting CO_2

The difference between the free carbon dioxide content and the associated free carbon dioxide produces the excess free carbon dioxide, which is also called the rust-promoting carbonic acid (Figure 10-4).

According to this, an ideal cooling water should contain no excess free carbon dioxide, but only the quantity of associated free carbon dioxide required as a result of the equilibrium reaction (1).

Figure 10-5 provides a graphical representation of the calcium compound-carbonic acid equilibrium in accordance with equation (2). It shows how much free carbon dioxide corresponds to a particular quantity of carbonic acid present as carbonate hardness.

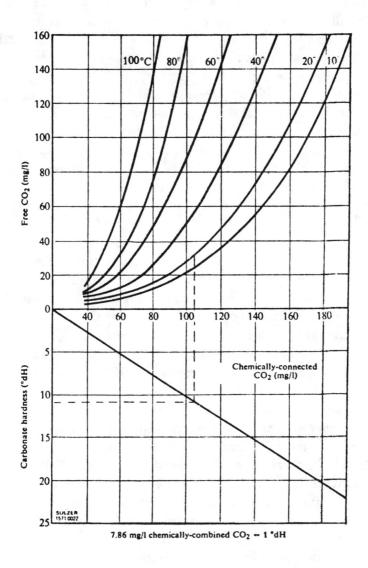

7.86 mg/l chemically-combined CO_2 = 1 °dH

Figure 10-5. Calcium Compound-Carbonic Acid Equilibrium with Consideration of the Various Water Temperatures

Chemical analytical determination of the total amount of carbonic acid in the water therefore enables us to ascertain whether the water will react aggressively, whether it will tend to precipitate a calcium compound or whether it can be considered as being a so-called equilibrium water.

Cooling waters without excess free carbon dioxide lead to the formation of a protective carbonate-rust scale which, in cooling systems, represents one of the most important protective covering layers. As indicated in Figure 10-6, it is created by including ferric hydroxide in the calcium carbonate layer which is formed through the reaction of the calcium bicarbonate with the hydroxyl ions that develop on the cathodic areas as a result of the corrosion of iron. The calcium carbonate can therefore be regarded as being a very effective cathodic corrosion inhibitor. With low water temperatures and the assistance of the dissolved oxygen, it leads to the creation of a thin impervious protective coating. As no further hydroxyl ions are produced and the transportation of oxygen to the cathodic areas is rendered more difficult once a certain coating thickness is attained or the cathodic areas are masked, both the corrosion of iron and the growth of the carbonate-rust protecting layer come to a standstill.

With rising temperatures, increasing quantities of associated free carbon dioxide are required to maintain a specified quantity of calcium bicarbonate in solution (Figure 10-5). In accordance with equation (1), this means that insoluble calcium bicarbonate is precipitated. One must therefore assume that the formation of a protective layer would be accelerated with a rise in temperature. This is not the case, however, as during temperature rise the approximation to the equilibrium and thus the calcium compound precipitation takes place too quickly and without regard to the readiness of the metal surface. Protective coatings are only created if the calcium compound is precipitated as a result of the alkalization of the metal surface. A thermal equilibrium disturbance occurs when water is heated up and the precipitated calcium carbonate—whereby this precipitation is not necessarily made onto the surface of the metal—leads to sludge formation and with higher temperatures to calcinated scale. The layers formed in this way, the thicknesses of which continue to increase, are mostly very thick, uneven and may only be regarded as protective coatings in exceptional cases.

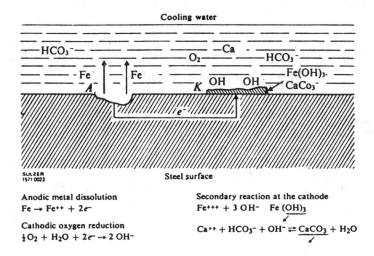

Cooling water

SULZER
1571 0023   Steel surface

Anodic metal dissolution
$Fe \rightarrow Fe^{++} + 2e^-$

Cathodic oxygen reduction
$\frac{1}{2}O_2 + H_2O + 2e^- \rightarrow 2 OH^-$

Secondary reaction at the cathode
$Fe^{+++} + 3 OH^- \quad Fe(OH)_3$

$Ca^{++} + HCO_3^- + OH^- \rightleftharpoons CaCO_3 + H_2O$

Figure 10-6. Model Showing the Formation of a Protective Layer by the Anodic and Cathodic Partial Reactions

One very important conclusion to be drawn from this explanation is that the described calcium compound-carbonic acid equilibrium should only be considered for the protection of warm-water systems if the temperatures in the entire system are homogeneous and they always remain more or less constant. These two conditions are practically next to impossible for cooling systems. In the standstill condition and in the refrigeration parts of the system, one always has a water with insufficient hardness and the freed carbon dioxide remains partly or wholly dissolved in the water as an iron-aggressive carbonic acid. As the equilibrium, i.e. the re-solution of the calcium compound coatings only sets in slowly again, the process is practically irreversible.

Apart from the pH value and the hardness, further dissolved substances also have to be considered when classifying the aggressiveness of the cooling water. The nature of these substances will only be mentioned briefly here.

Chloride

A high chloride ion content is undesirable, owing to its attribute of easily destroying the created protective films on a large number of metals. A medium-hard water should therefore contain less than 100 ppm of chloride: the admissible chloride content of soft waters is even lower, e.g. less than 50 ppm at 5° dH. With even softer waters, the chloride concentration must be kept still lower (Figure 10-7). In the case of austenitic steels, which are subjected to simultaneous tensile stress, traces of chloride ions can already start stress-corrosion cracking.

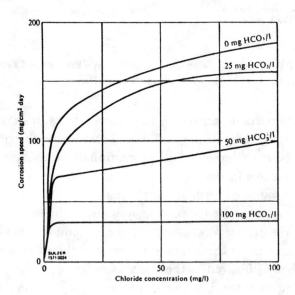

Figure 10-7. The Influence of the Cl-ion Concentration on the Corrosion Speed of Steel in Water with Various Degrees of Hardness (Ref. 5)

Sulphates

Although sulphates cause no increased corrosion to metallic materials, concrete cooling water basins are endangered as soon as the water contains more than 250 ppm of sulphate ions. This is because voluminous calcium-aluminum sulphates are formed which can cause the concrete to crack.

Ammonia

Ammonia or ammonia ions cause no excess corrosion on ferrous metals. Nevertheless, traces of this form of contamination will cause stress-corrosion cracking in brass alloys; copper and copper-nickel alloys are not susceptible to this type of corrosion, but a heavier and pronounced uniform corrosion will take place with higher ammonia contents.

Iron and Manganese

Less than 0.2 ppm of iron and manganese should be present. With higher contents, sludge-type hydroxides are precipitated in the presence of oxygen.

Suspended Matter

Suspended matter will only lead indirectly to corrosion damage. On the one hand, it may be responsible for the formation of aeration cells due to deposition onto the metal surfaces, this being a particular danger when excessively low flow velocities are involved ($<$1 m/s).

On the other hand, superpositioning of corrosion and erosion may take place and thus accelerate the corrosion of the metal. With higher suspended matter content of raw water, it is necessary to remove same by filtering, or to counteract the eroding effects by selecting suitable materials, e.g. aluminum bronze, 13 percent chromium steels, etc.

Dry Residue

Dry residue after evaporation includes all the dissolved substances in the water. If the salt content—expressed as dry residue—is greater than 500 ppm, a relatively good conductive water is at hand and galvanic corrosion can occur at the places of contact of different metals.

Dissolved Organic Matter

Dissolved organic matter is detected indirectly by determining the potassium permanganate consumption. If the $KMnO_4$ consumption exceeds about 25 mg/l water, then so-called "dirty" water is present which tends to produce sludge deposits and fouling.

Oxygen Content

The oxygen content should never be too low unless complete aeration is possible and an eventual oxygen uptake can be prevented. A minimum quantity of oxygen is needed to form the aforementioned carbonate-rust protecting film (Figure 10-6). This should amount to about 5-6 ppm of normal ambient temperature. On the other hand, cooling water saturated with oxygen will lead to increased corrosion.

Table 10-6 represents an example of a standard analysis for "good" cooling water. At the same time, this table shows which data are required to enable us to judge the quality of a cooling water.

The following may be stated in regard to the problematics of chemical cooling water analysis:

- Some contaminations—particularly the pH value as well as the dissolved oxygen and carbon dioxide contents—should be determined at the sampling location.
- It is often misleading to judge a cooling water on the basis of a single analysis. Whenever possible, judgment should be given

over a longer control period, e.g. a year. Only then is it possible to account for the periodic fluctuations in the quality of the water.

- With closed cooling circuits, it is necessary to supervise the cooling water quality constantly. The salt content and also the pH value may change slowly due to evaporation and the absorption of contaminations from the ambient air (e.g. dust, sulfur monoxide from combustion gases).

Table 10-6. Example of a "Good" Cooling Water
(Standard Values for a Nonaggressive Equilibrium Water)

Total hardness	17–19 °fH
Carbonate hardness	12–15 °fH
Non-carbonate hardness	2 °fH
Reaction at 20 °C	7.6–7.8 pH
Iron	< 0.2 mg/l
Chloride	< 50 mg/l
Sulphates	< 150 mg/l
Ammonia ions	< 5 mg/l
Free carbon dioxide	5–15 mg/l
Associated carbon dioxide	5–15 mg/l
Rust-promoting CO_2	0.0 mg/l
Oxygen	> 5 mg/l
Residue after evaporation at 105 °C	< 500 mg/l
Organic matter ($KMnO_4$ consumption)	< 25 mg/l

Sea and Brackish Water

These two cooling liquids will only be briefly mentioned at this point for the purpose of completeness. As a result of their high chloride content, sea and brackish waters are considered as being aggressive cooling mediums. Therefore, special materials are mostly required when handling them.

Treated Water

When treating water, we try to extract the disturbing contaminations by some chemical or physical reactions.

The purpose of the treatment is often only to remove the hardening agent from the water (softening); in certain cases it is advantageous to remove not only the hardening agents (calcium and magnesium ions), but all the dissolved salts (partial or complete demineralization).

Such treatments can be subdivided into three main groups:

- Precipitation treatment, by which the hardening agent is precipitated through the adding of chemicals. The residual hardness is of the order of about 1–4° dH (lime-soda process, hydrated-lime process, tribasic-sodium phosphate process, baryta process, etc.). The softened water reacts very alkaline.

- Softening by means of base exchange. This refers to the zeolites (alkaline earth or alkali-aluminum silicates) which are present in their sodium form and over which the water to be softened is led, whereby the calcium and manganese ions are replaced by the sodium ions of the zeolites (Figure 10-8). With this method, it is possible to reduce the water hardness to approximately 0.05° dH, whereby the water is not completely free of salt.

- On the other hand, complete demineralization is made with synthetic organic resins which enable the metal cations in the water to be replaced by hydrogen ions (cation exchange) and the anions by hydroxyl ions (anion exchange) (Figure 10-8). Water treated in this manner is salt-free and has a neutral reaction.

Soft and very soft waters produced by means of the described processes cannot form protective films on the surface of cast iron and unalloyed steel; they can therefore be considered as being extremely aggressive. A softening treatment should therefore only be considered when

- the water can be completely degasified to remove O_2, CO_2 or
- chemically inhibited.
- If such measures are not possible, particularly resistant metals should be employed (e.g. 13 percent chrome steels, austenitic Cr/Ni-steels, titanium, plastics and plastic coatings).

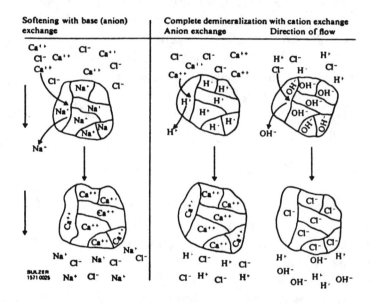

Figure 10-8. Schematic Diagram Showing the Reactions When Water is Softened by Means of Base Exchange *(left)* and Complete Demineralization Using Cation and Anion Exchanges *(right)*

Cooling Media

Cooling media are used in refrigeration and air conditioning systems to remove heat calories. There are two types of cooling media:

- Organic coolants (ethylene glycol-water or methanol-water mixtures). They are not only used as cooling media, but also as frost-resistant cooling liquids.

- Chloride cooling brines (concentrated aqueous brine solutions).

Water-Alcohol Mixtures

Pure alcohols (methanol, ethanol, glycerol, ethylene glycol) have no corrosive effect on metals—particularly on ferrous material. The same applies to alcohol-water mixtures, providing the mixtures are free of oxygen. If they come into contact with air, however, these mixtures will corrode unalloyed steels as soon as the water content exceeds 25 percent. The following metals are also susceptible to attack under these conditions: aluminum, cast iron, brass. The addition of a suitable inhibitor (e.g. borax or a mixture of Na_2HPO_4 and NaH_2PO_4) is absolutely essential. Chromates should never be used as inhibitor in this case.

Chloride Cooling Brines

Chloride cooling brines are used a great deal in the refrigeration industry because of their stability and cheapness. Sodium chloride (up to 22 percent NaCl) and calcium chloride solutions (up to 29 percent $CaCl_2$) are the most commonly used.

These chloride brines are good electrolytes and therefore have an extremely corrosive effect. The corrosiveness of these brines increases intensively when the pH value falls and the oxygen content rises. Consequently, it is customary to increase the pH value from 9.0 to 9.5 by adding sodium or calcium hydroxide. The pH value should not exceed 10. It is absolutely necessary to control the brine at regular intervals and correct the pH value in accordance with the respective requirements.

Generally speaking, we can say that $CaCl_2$ brines are less aggressive than NaCl ones, firstly because $CaCl_2$ is not so strongly dissociated in water as NaCl and, secondly, because the calcium ions promote the formation of protective films on steel. On the other hand, we should like to emphasize that the pH value of a $CaCl_2$ brine is not so stable as a NaCl one. As a result of the reaction with the carbon dioxide in the air, free hydrochloric acid is formed according to the equation.

$$CaCl_2 + CO_2 + H_2O \rightarrow CaCO_3 \downarrow + 2\ HCl$$

The pH value can fall from 8.5 to 5.5 within a few weeks—in smaller systems, in the matter of a few days.

It is therefore advantageous to adjust the pH value of a $CaCl_2$ brine with $Ca(OH_2)$ and not with NaOH, because HCl cannot be formed, at least as long as an excess of $Ca(OH)_2$ is present:

$$CaCl_2 + CO_2 + H_2O \rightarrow CaCO_3 \downarrow + 2\ HCl$$

$$2\ HCl + Ca(OH)_2 \rightarrow CaCl_2 + 2\ H_2O$$

With systems that use sodium chloride brines, the absorption of CO_2 tends to bring about a rather slow increase in the pH value (Ref. 2) and this has to be compensated by adding hydrochloric acid.

The aggressiveness of chloride cooling brines can be greatly reduced by the addition of corrosion inhibitors (Ref. 3). Nitrites and chromates are particularly suitable, but such inhibitors should not be used for foodstuff refrigeration because of their toxicity. In such cases, more expensive organic inhibitors like sodium benzoate have to be used.

It should also be pointed out that the danger of galvanic corrosion is particularly great because cooling brines are good electrolytes. For example, damage to welded steel pipelines has been observed which was only due to slight difference in the composition of the filling metal and the raw material.

General Aspects of Corrosion in Cooling Systems

As metal corrosion occurs in practically every medium, it seems advisable at this point to provide some information on the extent and type of corrosion to be expected. It is impossible to avoid corrosion in any system. The corrosion expert's task is to fulfill the following: to select the constructional material and the composition of the cooling medium in such a way that the corrosion remains within economic limits, and certain particularly dangerous types of corrosion are avoided.

A detailed description of these corrosion phenomena and theri respective causes would exceed the scope of this chapter. Nevertheless, attention is drawn to the data in Table 10-7. Here, the most important types of corrosion are related to the material attacked in each case, and the respective causes as well as the possible countermeasures to be taken are also shown. Two of the forms of corrosion enumerated will be discussed in somewhat more detail:

Galvanic Corrosion

Galvanic corrosion occurs when two different metals (i.e. two metals with different standard potentials) are conductively connected. As a rule, the less noble of the two solutionizes quicker than it would if exposed to the respective medium alone.

Such metal pairings are all the more dangerous the greater the distance between the individual standard potentials. Furthermore, the conductivity of the medium is also important: the danger of galvanic corrosion is enhanced with increasing conductivity.

Finally, it should be pointed out that the surface area ratio—noble to less-noble—also plays an important role: if the surface area of the less-noble metal is small, a high anodic corrosion current density is produced and local metal solutioning is greatly accelerated. When the conditions are reversed, i.e. the less-noble surface area is larger than the noble one, a relatively low anodic corrosion current density is created so that solutioning of the metal is only slightly accelerated. The metal pairings shown in Figure 10-9 as well as the galvanic corrosions produced in aqueous electrolytic solutions are based on whether the surface area relationship between the noble and less-noble metal is large or small.

Table 10-7. The Most Serious Forms of Corrosion in Cooling Systems

Type of corrosion	Originating agent or cause of damage	Material affected	Remedy
Uniform corrosion	In every cooling water	All	pH-adjustment, inhibition, choice of material
Pitting corrosion	Cl-ions	CrNi steel, Cr-steel, aluminium	Steel containing Mo, pure aluminium
Aeration cells	Suspended matter, air bubbles	Metals that can be passivated	Filtration, degassification, suitable construction
Stress-corrosion cracking	NH_3, NH_4^+ (ppm) Cl- (ppm fractions)	Brasses, austenitic steels	Choice of material (pure copper, CuNi), Steels with Ni > 25%, stainless chromium steels
Dezincification	Slight movement of water, Cl-ions, porous covering layers	Brass	Adequate flow velocities, choice of material: Brass + As or P, naval brass
Galvanic corrosion	Contact of different types of metal in electrolytic solutions	Dependent on the individual potentials	Avoidance of diverse pairings, constructive measures (insulation)
Corrosion by micro-organisms	Micro-organisms	Particularly Fe-metals	Chemical additives, strict adherence to operating instructions
Corrosion by flowing mediums	Too slow or fast running cooling water	All materials	Avoidance of speeds < 1.2 m/s, adjust max. speed to material used

Partner with smaller surface area

Partner with greater surface area	Au, Pt, Rh, Ag	Ni-Mo-alloys	Cu-Ni-Bronze	Cu, Brass	Ni	Pb, Sn	Fe, Carbon steel	Cd	Zn	Mg, Mg-alloys	Stainless steels	Cr	Ti	Al, Al-alloys
Au, Pt, Rh, Ag	▨	A	A	A	A	A	A	A	A	A	A	A	A	A
Ni-Mo-alloys	B	▨	A	A	A	A	A	A	A	A	A	A	A	A
Cu-Ni-Bronze	C	B	▨	A	A	A	A	A	A	A	B	B	B	A
Cu, Brass	C	B	B	▨	B	B	A	A	A	A	B	B	B	A
Ni	C	B	A	A	▨	A	A	A	A	A	B	B	B	A
Pb, Sn	C	B	B	B	B	▨	A	A	A	A	B	B	B	A
Fe, Carbon steel	C	C	C	C	C	C	▨	A	A	A	C	C	C	B
Cd	C	C	C	C	C	B	C	▨	A	A	C	C	C	B
Zn	C	C	C	C	C	B	C	B	▨	A	C	C	C	C
Mg, Mg-alloys	D	D	D	D	D	C	D	B	B	▨	C	C	C	B
Stainless steels	A	A	A	A	B	A	A	A	A	A	▨	A	A	A
Cr	A	A	A	A	A	A	A	A	A	A	A	▨	A	A
Ti	A	A	A	A	A	A	A	A	A	A	A	A	▨	A
Al, Al-alloys	D	C	D	D	C	B	B	A	A	A	B	B	C	▨

A   The corrosion of the supporting metal is not increased through contact with the metal X
B   The corrosion of the supporting metal is slightly increased
C   There is a pronounced increase in the corrosion of the supporting metal
D   This pairing should be avoided at all costs

SULZER
1571 0027

Figure 10-9. The Effect of Various Metal Pairings on Corrosion

Corrosion Due to Unfavorable Flow Velocities

Corrosion can be favored both by too low and too high cooling liquid flow velocities.

Low-flow velocities—below about 1 m/s—promote the deposition of solid substances onto the surface of metals: the aeration cells that are formed in such a manner often lead to heavy local corrosion. Consequently, critical conditions also exist when a system is shut down. Metallic materials behave quite differently when the flow velocities exceed 1 m/s: with metals that can be passivated (e.g. ferritic and austenitic stainless steels), we generally notice a reduction in the solutioning of the metal when the flow velocity rises. This is because the supply of oxygen is increased, thereby the passive films are strengthened and the existing defective film areas can quickly recover.

With unalloyed or low alloy steels, however, the quicker transportation of oxygen gradually accelerates the solutioning of the metal. The solutioning of the metal rises suddenly when a certain critical velocity, dependent on the type of metal and medium being used, is exceeded (Figure 10-10): apart from accelerated corrosion (no further adhesive protective films are formed), local metal destruction due to erosion also occurs in this range. Such high velocities should be avoided in a system. The following standard values apply in respect of the maximum admissible flow velocities:

Pure aluminium	1.2 m/s
Peraluman	1.5 m/s
Pure copper	1.8 m/s
Copper containing arsenic	2.1 m/s
Naval brasses	2.0–2.4 m/s
CuNiFe 90/10	3.0 m/s
CuNiFe 70/30	4.5 m/s
Steel	3–6 m/s
Nickel alloys	up to 30 m/s
Plastics	6–8 m/s

Deposition Phenomena in Cooling Systems

Water is an excellent solvent for numerous substances. It washes the mineral elements out of the earth and rock, gases are absorbed from the air (O_2, CO_2) and suspended matter is entrained. Furthermore, water constitutes a natural growth-promoting medium for microorganisms.

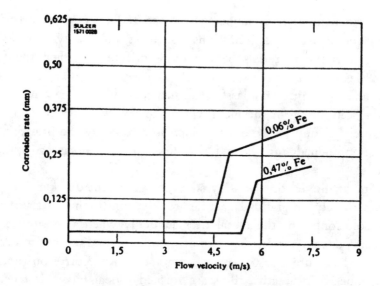

Figure 10-10. The Influence of the Flow Velocity of Sea Water on the Metal Loss on Copper-Nickel Tubes With and Without Iron Addition (Duration of Test = 60 Days)

If such cooling water is not treated, it leads to:

- the formation of mineral layers, often referred to as "scale" ($CaCO_3$, $CaSO_4$, $MgSO_4$). This scale hampers the heat transfer and increases the pressure drop in a heat exchanger;

- depositions of organic nature which—exactly like the scale—impair the heat transfer or cause pitting;

- irregular corrosion attack.

The formation of mineral layers is often described as "scaling," whereas organic films are known as "fouling."

The effect of such deposits on the heat transfer of a heat exchanger can be seen quite clearly in Table 10-8.

In addition to the intensely reduced heat transfer, contaminated cooling systems often call for up to 50 percent more pump motor output. The possibility of counteracting the reduction in output by overdimensioning the exchange surfaces can never be seriously considered in the light of the above data. The cooling water would definitely have to be treated.

Table 10-8. The Influence of the Carbonate Scale on the Heat Exchanger Capacity

Scale thickness (mm) [1]	Coefficient of total heat transfer [2]	Increase of heat-exchange surface (%)
0.000	850	0
0.130	595	45
0.260	460	85
0.520	315	170
1.040	240	250

[1] Assumed λ for the carbonate deposit: 1.0 Btu/hr/sq · ft/°F/ft
[2] Btu/hr/sq · ft/°F. Temp-Diff/sq · ft

Cooling System Design and Treatment Possibilities

In principle, damage to cooling systems can be checked in two different ways:

- By treating the corrosive medium, it can be changed in such a way that its aggressiveness is reduced; to this end, one can either remove the aggressive components to a great extent or add specifically-acting chemicals to the medium. In this way, both corrosion and deposition are avoided.

- By selecting a material which can be exposed to the aggressive medium without the danger of corrosion, or to change or influence the surfaces of the metal coming into contact with the medium in such a way (e.g. painting, cathodic protection) that corrosion is reduced to a minimum. The formation of deposits, however, is not significantly influenced by this treatment.

The treatment to be effected always depends on the cooling system design. In principle, one differentiates between three design types (Figure 10-11).

- Completely closed cycles
- Closed cycles with cooling towers
- Once-through cooling systems with direct cooling, whereby the water is taken from a river, for example, passed through the system and returned to the river again.

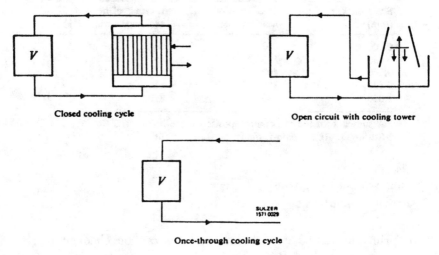

Closed cooling cycle

Open circuit with cooling tower

Once-through cooling cycle

Figure 10-11. Three Different Possibilities for the Design of a Cooling System

Completely Closed Cycles

The completely closed cycle represents the ideal possibility for clean and effective water treatment.

Filling the system with partially softened or even completely demineralized water is economically justified even in larger plants. The danger of incrustation is thereby avoided. Usually such systems provide no difficulties in respect of algae growth.

An inhibition must be undertaken, however, due to the low degree of hardness. Nevertheless, this can be made quite simply and cheaply. The make-up water required to compensate the low evaporation losses also has to be partly softened. If, however, the water loss is attributable solely to evaporation, then an inhibitor is not to be added to the make-up water; on the other hand, should a leakage occur then the specified inhibitor quantity must be added. The inhibitor must be checked periodically and readjusted whenever necessary.

Closed Cooling Cycles with Open Cooling Towers

The closed cooling cycle with open cooling towers presents a number of very difficult problems.

Large quantities of fresh water are required because of the high evaporation losses. Consequently, the preparation of softened or demineralized fresh water is often problematic. On the other hand, the use of a hard make-up water calls for relatively frequent desludging. Troublesome calcium compound deposits are found in the warmer parts of the system.

Thus, one endeavors to use a soft water as fresh water wherever possible. Furthermore, efforts are made to stabilize the carbonates so that they remain in solution or prevent the adhesion of precipitated carbonates on the walls of the system by adding suitable chemicals.

The intensive contact between the cooling water and air as well as the solar radiation stimulates the growth of algae and microorganisms. As a result, frequently algae destructive additives have to be added to the water as well, particularly in the warmer months of the year. The dead algae and microorganisms remain as extremely fine solid particles in the water and enhance the danger of increased deposition and incrustation. This can be effectively counteracted by adding dispersing agents which keep these solid particles suspended and prevent them from coagulating and forming deposited layers.

Furthermore, the water of such cooling systems should be provided with a corrosion inhibitor, because the formation of protective carbonate-rust films is deliberately hindered.

It should also be pointed out that the various additives are only effective over a very limited pH range. This means the respective pH value in the system should be strictly controlled and regulated by adding an acid (H_2SO_4) or a base (NaOH). Figure 10-12 shows the schematic arrangement of an automatically controlled plant for the admixture of chemicals in the return-flow basin of a cooling tower cooling system. The pH value is continually controlled and regulated, whereas the chemicals are added intermittently depending on the fresh water supply which is controlled by level detectors.

The fresh water used to compensate the evaporation losses requires no further treatment when very soft water is available. If, however, the cooling water is of medium-hard quality, the water must be given an additive to stabilize or disassociate the hardening constituents (additives on a phosphate basis, complex salts of organic compounds). Depending on the type of additive used, an increase in the concentration u_c of 2 to 5 is permitted. u_c represents the relationship between the evaporated water quantity Qv and the desludging Q_A.

$$u_c = \frac{Qv}{Q_A} = \frac{C_A - Cz}{Cz}$$

C_A Maximum possible concentration in the system
Cz Concentration in the make-up water

This value must be maintained, otherwise the maximum admissible salt content of the cooling water will be exceeded. The supervision of the increase in concentration, for example, is made by determining the Cl content of the circulating cooling water. The maximum admissible salt content must be specified according to the raw water quality and from case to case. According to the VGB directives (Ref. 4), the maximum total salt content of the cooling water should not exceed 3000 mg/l. This is not because of plant-operating reasons (e.g. corrosion depositioning), however, but to prevent damage to vegetation in the vicinity of the tower as a result of salt spraying.

With medium-hard to very-hard waters, it is necessary to remove some of the hardening constituents before introducing the fresh water into the cooling circuit. This can be effected with one of the following methods:

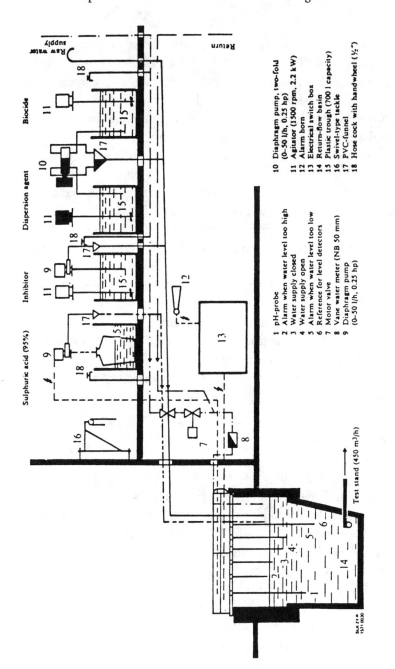

Figure 10-12. Schematic Arrangement of an Automatically Controlled Plant for the Addition of Chemicals in the Return-Flow Basin of a Cooling Tower

1 pH-probe
2 Alarm when water level too high
3 Water supply closed
4 Water supply open
5 Alarm when water level too low
6 Reference for level detectors
7 Motor valve
8 Vane water meter (NB 50 mm)
9 Diaphragm pump (0–50 l/h, 0.25 hp)

10 Diaphragm pump, two-fold (0–50 l/h, 0.25 hp)
11 Agitator (1500 rpm, 2.2 kW)
12 Alarm horn
13 Electrical switch box
14 Return-flow basin
15 Plastic trough (700 l capacity)
16 Swivel-type tackle
17 PVC-funnel
18 Hose cock with handwheel (½")

- Precipitation of the hardening constituents by adding lime water. This method produces a very alkaline, almost hardness-free water whose salt content is reduced by the amount of carbonate hardness. It is particularly suitable for plants with high fresh water consumption. Considerable quantities of sulfuric acid are required for neutralization purposes.
- Softening by means of the base exchange process produces a hardness-free water. As the calcium and magnesium ions are replaced by sodium ones, the salt content is not reduced to any great extent. This method is only suitable for small installations where considerable desludging losses can be accepted.
- Softening and decarbonating over exchange resins, whereby the hardening constituents and the associated carbonic acid are removed from the water. The salt content of the water is reduced by the amount of carbonate hardness. This method is suitable for small- to medium-sized plants.
- Complete demineralization is only used for specific purposes because of the increased costs. Generally speaking it is only considered for installations where a demineralization plant already exists.

Once-Through System

The third type of cooling system is the once-through process. This is used when extensive and cheap supplies of water are available (e.g. river-water cooling). Although the plant costs are at a minimum with such cooling systems, the possibilities available for water treatment sre practically nil. Only filtration of the incoming raw water and innoculation with dispersing agents can be considered. The adding of other possibly more-effective chemicals is mostly ruled out due to economic reasons or the danger of sewage pollution. The primary object of water treatment in such plants is limited to keeping the metal surfaces free of deposits and contamination.

Protection Against Corrosion

Protection against corrosion can be aimed at in two ways, namely by adding a corrosion inhibitor or using cathode corrosion protection.

Corrosion Inhibitors

The main requirements placed on an inhibitor are:
- Cheap and effective even in small concentrations (mostly 0.5–1%).
- Consistent corrosion protection even with excessive and particularly, under dosages.
- No sewage problems are to be raised through the use of an inhibitor.

The most important inhibitors are nitrites (possibly combined with borates), chromates, organic substances (such as tannin, amine, benzoates), zinc phosphate and zinc/chromate phosphate.

The following inhibitors are less effective: silicates (ineffective with water speeds exceeding 2 m/s), polyphosphates, soluble oils (the emulsion are easily destroyed by excessive salt content, dirt or very high temperatures).

The products available on the market are usually made up of the enumerated chemical products or mixtures of same.

Cathodic Corrosion Protection

Cathodic corrosion protection can either be provided through the fitting of sacrificial anodes (magnesium, zinc) or by connecting the plant to an external source of power. In the former case, the equipment to be protected is brought into electrical contact with a less-noble metal (Figure 10-13), which is selectively solutionized, whereas the equipment plays the role of a cathode and therefore does not corrode. If, however, the protection is afforded by means of impressed current, the equipment is connected to the negative terminal of the power source, whereby its potential is displaced in the cathodic range. Consequently, corrosion is practically impossible as long as the current density needed for cathodic protection is maintained (Figure 10-14).

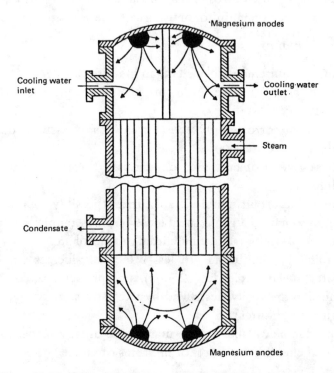

Figure 10-13. Schematic Representation of the Magnesium Anode Arrangement for the Protection of the Tube Plates of a Heat Exchanger

The cathodic protection of heat exchangers certainly allows the water chambers and the ends of the heat exchanger tubes to be protected, but not the tube interiors. Following economic considerations, we can reply to the question as to whether the protection should be provided by means of sacrificial anodes or impressed current to the following manner: with low power consumption, protection through sacrificial anodes is cheaper, whereas with higher power consumption (from 5 A per year) it is advisable to employ impressed current.

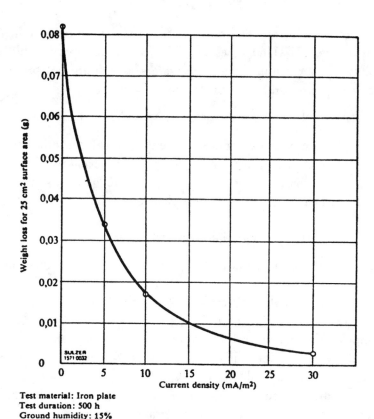

Figure 10-14. The Effectiveness of the Cathodic Protection Against Ground-Level Corrosion in Relation to the Protecting Current Density

To keep the solutionizing of the sacrificial anodes and the power consumption within reasonable limits, the metal surfaces should be provided with a protective coating (paint). In this way, the cathodic current only has to protect the damaged areas of the protective coating. The selected coating should be sufficiently alkali-proof (e.g. up to pH 11) because the formation of hydroxyl ions is increased in the cathodic areas. With cathodic protection, the cathodic reaction occurring in neutral waters

$$O_2 + 2 H_2O + 4e^- \rightleftarrows 4 OH^-$$

is transferred onto the steel surfaces that have to be protected. An increase in the pH value takes place on the metal surface and the precipitation of calcium from such waters, which are not prone to calcium compound precipitation, is encouraged:

$$Ca^{++} + 2\ HCO_3^- + 2\ OH^- \rightleftarrows CaCO_3 + CO_3^{--} + 2\ H_2O$$

It is therefore essential to reduce the density of the cathodic protective current once the natural carbonate rust-protecting film has been built up. Otherwise, calcium compound and sludge deposits on the protected surfaces will interfere with plant operations.

Extermination of Algae and Microorganisms

The following treatments are employed for the extermination of algae and microorganisms:

- Chlorination with addition of 1–2 g of chlorine per cubic metre. The residual chlorine content should not exceed $0.1\ g/m^3$.
- Shock chlorination, by which up to 10 g chlorine/m^3 is added to the water periodically at intervals of 2–3 minutes. This form of treatment may enhance corrosion.
- The adding of copper salts (several grams of copper per cubic metre) is less effective because certain microorganisms become immune to the treatment with time.
- Treatment with chlorinated phenols has not been successful because of the disturbing smell.
- Quaternary ammonium compounds are effective additives and they are widely used today despite their relatively high cost.
- Despite its great advantages, the ozonization of cooling water can hardly be considered owing to the high costs involved.

Dispersing Agents

The task of a dispersing agent in cooling water is to prevent the depositioning of suspended solid particles, particularly onto the heat exchanger tubes. Furthermore, the sludge present in the cooling water and the salts of low solubility precipitated from the latter are kept finely dispersed and suspended in the cooling water or removed by means of desludging or in the by-pass filter. These additives can also be used for cleaning contaminated systems under operational conditions.

There are natural and synthetic dispersing agents. For example, natural dispersing agents are certain derivatives of lignin sulfonates which are produced during the manufacture of paper. Synthetic dispersing agents are uncharged hydrophilic polymers and anionic or cationic-charged polyelectrolytes (e.g. sodium polymethacrylate and polyvinyl pyridinium butyl bromide).

Whereas the manner of operation still remains unclear in the case of the uncharged polymers, the polyelectrolytes act by depositing electrically-charged particles onto the contaminants. With weaker concentrations (several grams per cubic metre), this leads to flocculation and, in the case of higher concentrations (25–300 g/m^3), to dispersion. The manner of operation is schematically illustrated in Figure 10-15.

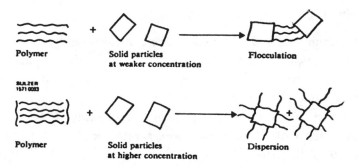

Figure 10-15. Action of Dispersing Agents at Various Concentrations (Dispersion and Flocculation) (Ref. 6)

Stabilizers

On reaching the solubility limit, stabilizers are able to retard the nucleation of individual compounds of low solubility, and prevent any existing crystals from forming adhesive deposits.

The manner of operation can differ:

- The agent can cover the crystal surface and, by changing the physical surface quality, avoid the cohesion of the crystals with each other and with the surface of the metal.
- The agent can retard the speed of crystalline growth due to absorption on the surface of the crystal.

Material Selection—Economic Considerations—Constructive Measures

The question whether corrosion and depositioning in cooling systems can be checked by means of suitable material selection or water treatment must be answered from case to case on the basis of corrosion-chemical and economic considerations.

When selecting the material, one always endeavors to use the cheapest possible material. As a result, we try to employ unalloyed cast iron and steel in the first place. Alloyed iron material (e.g. austenitic Ni-cast iron, stainless chromium steel or austenitic steel or nonferrous metals [copper, brass, CuNi]) are only considered in special cases.

Aluminum and Aluminum Alloys

Aluminum and aluminum alloys are being increasingly used for the construction of heat exchangers. The use of aluminum material may only be recommended, however, when the cooling water chemistry can be completely controlled.

It is particularly important to keep the surfaces of the metal free of every form of deposit; otherwise aeration cells with pitting-type corrosion will form. Aluminum should not be used in weak acid solutions (pH value <6) or alkaline waters (pH value >8.5).

Plastic Coatings

Plastic coatings often represent the ideal solution to a cooling water problem. Under certain conditions, they allow a cheap and corrosion-susceptible metal to be covered with a high-grade and resistant plastic coating.

To ensure satisfactory protection, these coatings—polyethylene, PVC, epoxy resins, polyamides—must be nonporous and of sufficient thickness (150–250 μm).

In the case of plastic coatings, the low thermal conductivity is often regarded as being a disadvantage. Actually, it only amounts to $\lambda \cong 2$ kcal/m $\cdot$ h $\cdot$ °C against 39 kcal/m $\cdot$ h $\cdot$ °C for steel. This disadvantage is only fictitious, however, because no heat transfer retarding films are formed from corrosion products on plastics, as is the case with most metals. After a short period of operation, the thermal conductivity of a steel tube coated with plastic is therefore superior to one without.

As plastics are chemically resistant and some of them have nonhydroscopic properties as well, depositions are formed in exceptional cases only; the pressure drop in the equipment remains almost constant.

Some duroplastic coatings, which have been especially developed for this purpose, are suitable for temperatures up to 100°C or more.

We often have to contend with already existing cooling systems when the cooling water treatment has to be determined. It has been pointed out that an inhibition is certainly feasible in the case of a closed circuit, whereas with a once-through system, we are limited to a more or less effective innoculation. In the first case, one has an optimum solution, in the second, only a temporary one.

As far as cooling tower plants are concerned, the projected treatment chiefly depends on the fresh water consumption, the raw water price and the sewage charges.

Where raw water prices are high, partial demineralization is recommended, whereas with medium and low prices, it is advantageous to use calcium hydrate plants and chemical water treatment respectively. Table 10-8 provides a summary of the plant and operating costs for various water treatment processes. It is based on a fresh water consumption of 5 m³/h with a carbonate hardness of 28 °fH.

From the statements made so far, it follows that the solving of cooling water problems must be a part of the plant planning operation.

Table 10-8. Comparative Summary of Costs Incurred for Treating 5 m³ of Fresh Water an Hour with a Raw-Water Quality of CH = 28 °fH and TH = 30 °fH

	Calcium hydrate plant	Base exchanger	Partial demineralization	Total demineralization	Chemical treatment
Water quality	TH: 5-6 °fH CH: 3-4 °fH RS: 70-80% pH: strongly basic	TH: 0.1 °fH RS: unchanged (0%) pH: unchanged	TH: 0.1 °fH RS: 94% pH: unchanged	TH: 0 °fH RS: 100% pH: 7	TH: unchanged RS: unchanged pH: unchanged
Plant price (complete)	about Sfr. 30000.-	Two-fold plant about Sfr. 12000.-	Two-fold plant about Sfr. 45000.-	Two-fold plant a) manual operation about Sfr. 110000.- b) fully-automatic about Sfr. 175000.-	about Sfr. 20000.-
Cost of materials and chemicals[1]	8-10 Rp./m³	11 Rp./m³	20 Rp./m³	30 Rp./m³	15 Rp./m³
Attendance	2-4 h/day	½-1 h/day	1-2 h/day	3-4 h/day	1-2 h/week
Application	Make-up water for cooling tower, combined with chemical treatment	Make-up for closed circuits	Make-up for cooling tower, combined with chemical treatment	Special usage combined with chemical treatment (too expensive for cooling tower)	Cooling tower operation, ideal with relatively soft water

[1] 100 Rp. = 1 Sfr.

TH Total hardness, CH Carbonate hardness, RS Reduction of the total salt content

Furthermore, there is also a series of guiding principles to be considered when projecting the plant onto the drawing board. Here are a few useful directives:

- Care should be taken that the cooling system can be completely emptied when in the standstill condition. Numerous cases of corrosion occurring in cooling systems are attributable to standstill corrosion. If the system is not emptied, then one must have the possibility of recirculating the contents several times a day.
- When welding on the cooling water side, care should be taken that protruding weldments or crevices, as a result of unsatisfactory weld penetration, are avoided.
- The sealings should be designed in such a way that they cannot lift on the water side.
- The pairing of very different materials should be avoided. Under certain conditions, electrical insulations must be inserted.
- Excessive fluid motion should be avoided.
- The cooling water return lines should discharge below the surface level of the water.
- No air is to be taken up by the pumps as a result of defective stuffing boxes.
- The struts of vessels should be arranged externally and not internally.
- Narrow gaps should be avoided in the design stage. If this is not possible, the gap width should be preferably large (<0.5 mm).
- The flow velocities should be neither too low nor too high.

By considering these few, but extremely important, guidance principles on the one hand, and through the use of carefully planned water treatment on the other, it should be possible to reduce the difficulties in cooling systems to a minimum.

Bibliography

1. Tillmans, J: Z. Unters. Lebensmittel, Vol. 58 (1929), p. 33–52.
2. Piatti, L., Fot, E, Weber, J: Werkstoffe u. Korrosion, Vol. 18 (1967), p. 214–217.
3. Piatti, L., Fot, E: Werkstoffe u. Korrosion, Vol. 13 (1962), p. 597–66.

4. Richtlinien für die Aufbereitung von Kesselspeisewasser und Külwasser, VGB, 5th edition, Essen 1958, p. 160.
5. Held, H. D.: Mitteilungen d. VGB, 1965, No. 96, pp. 161–174.
6. Held, H. D.: Techn, Überwachung, Vol. 7 (1966), No. 6, p. 194.

How to Control Bio-Slime in Condenser Cooling System Water

If slime fouling is a problem which must be controlled in your cooling water system, chlorine is probably the first answer that comes to mind. This is not surprising since chlorine, with its low cost per pound, has enjoyed widespread use as a biocontrol compound. However, today a growing number of utilities are finding that chlorine is no longer always the best choice.

Current federal regulations limit the free available chlorine residual discharged in utility once-through cooling water or cooling tower blowdown to a maximum of 0.5 mg/l and an average of 0.2 mg/l. Chlorine residuals may not be discharged from any unit for more than two hours per day or from more than one unit at a time. Individual state regulations may be even stricter than these.

In view of current restrictions on chlorine residuals and with the possibility of tighter restrictions not far away, chlorine will undoubtedly become a less desirable biocontrol compound for utilities in the future. Because of this, many utilities have found it necessary to search for effective alternative biocontrol compounds in addition to chlorine. Fortunately, there are alternatives available that are already being used effectively for the power industry.

In addition to chlorine, other oxidizing biocides, nonoxidizing biocides, and surfactants are used for condenser biocontrol. A brief look into the capabilities of chlorine and its alternatives is presented in this chapter. This provides the utility engineer with a better perspective from which to choose the biocontrol program best suited for his particular system. An important consideration in evaluating cost performance of a biocontrol program is also discussed.

Oxidizing Biocides

Chlorine is commercially available either as a compressed gas or in the form of various chlorine-release compounds, such as hypochlorites or chloroisocyanurates. When chlorine is added to water, it reacts with water as follows:

$$Cl_2 + H_2O \rightarrow HOCl + HCl \qquad (3)$$

Hypochlorous acid (HOCl) is the most effective biocidal form of chlorine. In nonalkaline cooling water systems which are not subject to organic or ammonia contamination, hypochlorous acid is often a very effective and economical means of microbiological control.

In alkaline water, however, hypochlorous acid dissociates to the hypochlorite ion (OCl⁻) in increasing amounts, as demonstrated in Figure 10-16. The hypochlorite ion is 80 to 300 times less effective as a disinfectant than is hypochlorous acid.

Chlorine effectiveness also declines when it reacts with the ammonia-nitrogen present in many surface and reclaimed waters to form chloramines:

$$NH_3 \xrightarrow{Cl_2} NH_2Cl \xrightarrow{Cl_2} NHCl_2 \xrightarrow{Cl_2} NCl_3 \qquad (4)$$

The formation of mono-, di-, and trichloramines involves the substitution of chlorine for each hydrogen atom on the ammonia molecule. Between pH 7.0–8.0 the formation of dichloramines requires a weight ratio of chlorine to ammonia of 10:1. This really means that 10 mg/l of chlorine will be consumed in cooling water by 1 mg/l of ammonia. The resulting chloramines exhibit far less biocidal effectiveness than free chlorine (Figure 10-17).

Environmentally, chlorine has demonstrated aquatic toxicity at low levels. A 1975 study revealed a 94-hour TL_{50} of total residual chlorine (the limit that 50 percent of the test population could endure for 94 hours) to fish ranging from 0.08 to 0.26 mg/l. The same study found an invertebrate TL_{50} (94 hour) ranging from 0.21 to greater than 0.81 mg/l. Recent investigations into chlorination of

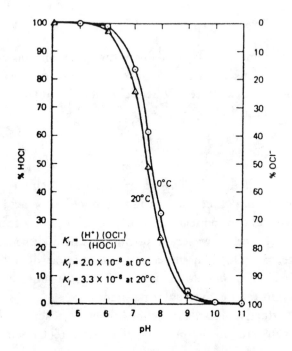

Figure 10-16. Dissociation of Hypochlorous Acid vs pH

drinking water have shown that chlorination of certain surface waters containing humic acids forms trihalomethanes (THMs). Concern with these compounds relates to possible carcinogenicity of THMs. For these reasons, federal and state agencies have imposed stringent restrictions on chlorine residuals in the cooling water effluent.

Chlorine dioxide (ClO_2) is a stronger and more selective oxidant than chlorine. It is currently used in well over 80 industrial and food processing facilities for cooling water biocontrol. Many of the larger plants use chlorine dioxide in several cooling systems. ClO_2 is used in a number of steam-electric utility cooling systems as well and its applications are expected to continue growing where an alternative to chlorine is required.

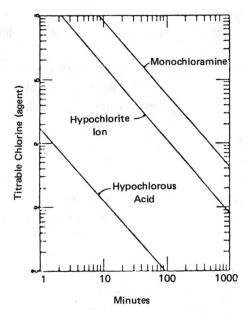

Minutes

99 per cent Distruction of E. Coli at 2-8°C

Figure 10-17. Germicidal Efficiency of Hypochlorous Acid

Chlorine dioxide is the product of reaction of hypochlorous acid with a suitable chlorite salt. ClO_2 is efficiently generated on-site in a two-step reaction process:

$$Cl_2 + H_2O \rightarrow HOCl + HCl \tag{5}$$

$$HOCl + HCl + 2NaClO_2 \rightarrow 2ClO_2 + 2NaCl + H_2O \tag{6}$$

Since the solubility of ClO_2 in water is 2.9 g/l at room temperature, reaching a solubility of over 10.0 g/l in chilled water, it is safely generated on-site in a water stream prior to injection into the condenser cooling water.

Basic on-site generation technology involves the use of diaphragm-type chemical feed pumps, packed reaction column, standard gas chlorinator and integrated safety controls. Since many utilities are already quite familiar with chlorination equipment, chlorine dioxide generation should involve very little unfamiliar equipment or procedures.

ClO_2 has several distinct advantages over chlorine which make it a desirable alternative. Unlike chlorine, the effectiveness of chlorine dioxide is relatively unchanged within the operating pH range of most condenser cooling waters (pH 6-9). Figures 10-18 and 10-19 compare the effectiveness of chlorine and chlorine dioxide against two slime-forming organisms at pH 7.0 and 9.5.

Of further importance, chlorine dioxide does not react with ammonia-nitrogen. Reaction (4), described earlier, does not take place. Therefore, when ammonia-nitrogen is present, ClO_2 can remain available to perform effectively as a microbiocide where chlorine may not. Chlorine dioxide does not form THMs in its reaction with the humic acids present in many surface water supplies. The net result is that chlorine dioxide can prove more effective than chlorine for microbiological control in many condenser cooling water systems. The following case histories illustrate actual field experience.

Case History 1: A West Coast utility initiated a program utilizing secondary treated sewage plant effluent as cooling tower make-up in 1967. This program presented a unique biocontrol challenge as the sewage plant effluent contained a fluctuating bacterial count and microbiological nutrients such as phosphate and ammonia. Concentrations of these nutrients in the cooling system further increased the problem of slime fouling.

Chlorine had always been the primary biocide in this utility cooling system. It had been supplemented by quaternary ammonium compounds, and earlier by chlorinated phenol biocides. Although the make-up water typically contained 2-10 mg/l combined chlorine, total plate counts in the cooling tower water were found to run as high as 500,000 organisms/ml. Rarely was a free chlorine residual present in the cooling water because of its reaction with various organics contained in the sewage effluent make-up water. After the start of this program, restrictions limiting the total chlorine in

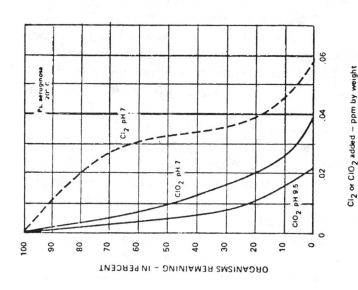

Comparison of Bactericidal Effects of Chlorine and Chlorine Dioxide on
A. aerogenes and *Ps. aeruginosa* at Different pH

Figure 10-19

Figure 10-18

the receiving stream to 0.2 mg/l were imposed. A more effective method of microbiological control had to be found.

Early in 1976, chlorine dioxide was suggested as a replacement for chlorine because of its biocidal effectiveness at low levels and its nature as a selective oxidant. On-site studies indicated that a feed of 1 mg/l of chlorine dioxide gave a trace to 0.2 mg/l residual in the cooling water.

A chlorine dixoide dosage of 1 mg/l was established at the point where sewage effluent enters as make-up. Since the initiation of the program, total plate counts have ranged from less than 10,000 organishs/ml down to 1,600 organisms/ml. Total coliform counts dropped from >2,400 organisms/100 ml to 100 organisms/100 ml with numerous tests of 10 organisms/100 ml.

In addition to improved microbiological control, corrosion rates have shown improvement. Mild steel and particularly admiralty corrosion rates were reduced, under the ClO_2 program, as measured by coupons and Corrater® probes.

Case History 2: An 880 MW capacity station with Mississippi River water for once-through cooling has used chlorine dioxide since October, 1976. Prior to that time, the utility controlled slime formation by chlorine injection at a dosage of 12 mg/l for 15 minutes per shift per condenser. Chlorine residuals varied, but ranged higher than current effluent regulations on free chlorine. Annual volume of chlorine consumed totaled 1.04 million pounds at a cost of nearly $91,250 at 1976 prices. Although chlorine was effective at this level, the utility was concerned with the environmental impact of this chlorine injection. An alternative approach was needed.

The utility chose to use chlorine dioxide on a trial basis since ClO_2 was a proven biocide over a wide pH range and did not react with ammonia-nitrogen present in the river water supply. Today, chlorine dioxide is still used at this station and continues to maintain effective control of condenser biofouling. The chlorine dioxide *dosage* is 0.2 mg/l, fed for 15 minutes per condenser, three times per day. Chlorine dioxide residuals typically range from a trace to

0.05 mg/l (expressed as ClO_2) which is equivalent to a trace to 0.026 mg/l (expressed as Cl). In 1978, costs for chlorine dioxide generation chemicals were *less* than the 1976 chlorine costs given above.

Nonoxidizing Biocides

Nonoxidizing biocides are generally organics of higher molecular weight than oxidizing biocides. They also tend to be slower acting than chlorine or chlorine dioxide, but can be effectively used alone or in conjunction with oxidizers for biocontrol where sufficient retention time is available. Minimal retention time, as well as cost considerations, typically prohibit the use of such compounds in once-through condenser systems. However, they are effectively used in open recirculating cooling systems. They include: quaternary ammonium compounds, chlorinated phenols, organosulfur compounds, and brominated organics.

Surfactants

Surface active compounds can greatly improve the performance of microbiological control programs. Surfactants enhance biocontrol programs by effectively reducing the surface tension when added to cooling waters. Alteration of the surface tension is typically achieved by the presence of a hydrophilic and a hydrophobic portion within the surfactant molecule. Water is attracted by the hydrophilic portion while a nonaqueous surface is attracted by the hydrophobic portion. These forces bring about decreased surface tension and allow the water to more effectively coat (or wet) the surface it contacts.

Since the cleansing ability of water depends on its ability to wet the surface to be cleaned, surfactants increase the mechanical cleansing action of the cooling water as it flows through the system. Organic (including microbiological) deposits may, therefore, be resuspended more readily in the system. The reduction in surface tension may also allow improved control of slime deposits by the primary biocontrol chemicals added to the system. The combined application of the surfactant at low levels in conjunction with an oxidizing biocide can prove very effective in improving microbiological control. Increasing use of surfactant technology will be used in the future to assist in minimizing the amount of biocide required for effective slime control.

Comparing Costs

The cost of a microbiological control program is an important concern. However, with a better understanding of the capabilities of the biocontrol chemicals used in the power industry, it becomes clear that cost per pound cannot be the sole basis for establishing a control program. Different biocontrol compounds meet different needs. Drawing from the choices available, individual control programs can be developed to solve individual biofouling problems. The program that provides the best control within existing restrictions can pay for itself in excess fuel or replacement power savings.

Think for a moment about the effect slime fouling of condenser tubes can have on utility costs. One utility calculated that a 0.12" Hg (absolute) increase in turbine back pressure on an 860 MW unit could cause a net loss of 0.1 percent on the heat rate. Resultant excess monthly fuel costs were calculated at $8,917. During the summer months, when condenser inlet water temperature increases, a 0.12" Hg increase over expected back pressure could result in a monthly excess fuel cost of $31,208. Comparing excess fuel (or replacement power) costs such as these to the cost of the biocontrol program provides a yardstick for evaluating cost-effectiveness of the program in use.

In summary, a number of oxidizing and nonoxidizing biocides are currently used for biocontrol. However, yesterday's solution to slime fouling problems may not apply today in view of tightening effluent restrictions on chlorine. Thus, selection of biocontrol compounds, cannot be made indiscriminately. In selecting alternative biocontrol compounds, consideration should be given not only to costs but to effectiveness of the biocide in each particular application, taking into consideration cooling water quality, type of microbiological fouling and discharge restrictions. Thoughtful selection of the biocontrol compound(s), alone or in combination with surface active agents, is essential for maintaining maximum microbiological control in condenser cooling water systems.

References

1. G. C. White, *Handbook of Chlorination*, Van Nostrand Reinhold Co. (1972).
2. G. C. White, *Disinfection of Wastewater and Water for Reuse*, Van Nostrand Reinhold Co. (1978).
3. A. A. Stevens et al., "Products of Chlorine Dioxide Treatment of Organic Materials in Water," EPA/IOI Workshop on Ozone/Chlorine Dioxide Oxidation Products of Organic Materials, Cincinnati, Ohio (November 17-19, 1976).
4. G. M. Ridenour and E. H. Armbruster, "Bactericidal Effect of Chlorine Dioxide on Some Common Water Pathogens and Sanitary Test Organisms," *J. Am. Water Works Assoc.* 41 (1949).
5. M. A. Benarde et al., "Efficiency of Chlorine Dioxide as a Bactericide," *Applied Microbiology and Biotechnology* (September 1965), Springer-Berlag, Berlin, Germany.
6. H. J. Gray and A. W. Speirs, "Chlorine Dioxide Use in Cooling Systems Using Sewage Effluent as Makeup," Cooling Tower Institute (January 23-25, 1978).
7. R. B. Thompson and J. A. Mathews, "Power Plant Operations and the Role of Condenser Tube Biofouling," Electric Power Research Institute Symposium, Atlanta (March, 1979).
8. R. N. Shreve and J. A. Brink, *Chemical Process Industries*, McGraw-Hill (1977).

Suggested Guidelines for Feedwater for Flash-Type and Oil-Field Boilers[1]

Feedwater-related problems in steam production equipment can generally be divided into scaling and corrosion factors. Scaling is the production of adherent solid deposits which produce localized overheating, poor heat transfer, and differential corrosion. The production of suspended solids or sludge in the cooler sections of the equipment is not considered scaling but is produced by the same

---

[1] Courtesy of Natco Boiler Co.

factors. Sludge production is undesirable in once-through oil-field steam generators since sludge carryover could produce plugging of the injection zone.

Corrosion is chemical attack on the boiler metal by oxidizing agents in the feedwater. Corrosion takes many forms and its rate may be modified by many factors which will not cause corrosion by themselves. Corrosion control consists of eliminating the oxidizing agents as much as practical and protecting the metal surface from the residual oxidants.

The following factors should be considered in the selection of feedwaters:

Total Hardness

The hardness ions, calcium and magnesium, are the most common causes of boiler scale. Many calcium and magnesium salts exhibit decreasing solubility with increasing temperature. This tendency leads to the precipitation of phosphates, sulfites, and carbonates which serve necessary functions in corrosion prevention.

For these reasons it is necessary to soften feedwater to one part per million (ppm) or less total hardness.

Total Dissolved Solids

Levels of total dissolved solids (TDS) in operation become a cause for concern only when liquid phase concentrations approach solubility limits. Therefore, a generator producing 80 percent quality steam should be able to tolerate feedwater salts in concentrations approaching 20 percent of their solubility limits. For example, sodium chloride levels of 60,000 ppm are possible. At extremely high TDS levels, control of corrosion-producing oxidants and scale-forming materials such as total hardness becomes critical.

The practical limitation on TDS generally comes as a result of water softener operating limitations. Most softening is done with resin ion exchange systems because of their reliability and simplicity of operation. Most softeners utilize salt regenerated sulfonate (strong acid) resins whose hardness leakage characteristics limit their operation to approximately 7000 ppm TDS. Recently developed weak acid resins are able to function at TDS levels up to approximately 30,000 ppm; however, both the capital and operating costs of these resins are several times that of the older resins.

Neutral Anions—Chloride, Sulfate

The concentration of these materials must be considered primarily in relation to their contribution to TDS. Since chloride is an effective depolarizing agent for corrosion cells, oxidant levels must be carefully restricted in the presence of large chloride concentrations.

Sulfate poses little problem except in the presence of calcium or barium with which it precipitates. Careful control of total hardness is necessary in the presence of large sulfate concentrations.

Heavy Metals—Iron, Manganese, Copper

These materials contribute to formation of complex mineral scales and must be controlled. Only iron is routinely found in natural waters; however, all three may be produced as corrosion products within the feedwater system. Ion exchange systems (softeners) are extremely efficient in the removal of heavy metals; therefore, heavy metal concentrations within the boiler are seldom a problem.

Heavy metals contribute greatly to fouling of ion exchange resins thereby necessitating expensive resin cleaning or replacement. Feedwater to the softeners should be restricted to 2 ppm or less of heavy metals.

Light Metals—Sodium, Potassium, Aluminum

Sodium and potassium salts are extremely soluble and need to be considered only in terms of their contribution to TDS. Aluminum is not a problem by itself but it can become a constituent in complex mineral scales. Soluble aluminum is seldom present in natural waters in significant amounts.

Suspended Solids

Suspended solids contribute to softener fouling and boiler sludge formation. In addition, chemical reactions of suspended solids inside the boiler can generate the components needed for complex mineral scale formation. Filtration of the feedwater to reduce the suspended solids levels is usually advisable. Suspended solids levels should be maintained below 5 ppm and preferably below 1 ppm.

Oil

The American Boiler Manufacturers Association guidelines call for oil contents of 7 ppm or below. Oil causes scale adherence, film boiling, and coking. Ion exchange resins are quite prone to oil fouling and tend to remove most feedwater oil at the expense of resin life. Maximum feedwater oil content of 5 ppm or less is recommended.

Oxygen

Oxygen is the primary corrosion accelerator in boiler water. Combinations of oxygen exclusion, mechanical deaeration, and chemical scavenging should be used to maintain oxygen levels as low as possible. Recommended maximum oxygen residuals range from 0.05 ppm to 0.007 ppm.

In practice, a constant level of scavenging chemical such as catalyzed sodium bisulfite is usually maintained.

Sulfite

Sulfite concentrations usually result from oxygen scavenging operations. Since calcium sulfite is an insoluble precipitate, calcium levels must be controlled when this form of oxygen removal is used.

Sulfide

Sulfide is unacceptable in boiler water because of the high rate and severe nature of the corrosion it produces.

Alkalinity

Alkalinity is primarily the result of carbonates in natural waters but may also be produced by hydroxides, phosphates, and sulfites. The function of alkalinity in an oil-field steam generator is quite different from that in a boiler producing dry steam. Although excess hydroxide alkalinity can contribute to caustic embrittlement, more moderate alkalinity levels help reduce corrosion and maintain silica solubility. Bicarbonate alkalinity levels of over 2000 ppm should be avoided due to excess hydroxide production in the boiler.

Vapor phase corrosion protection may be necessary in systems operating with high alkalinity or where separation of the steam and liquid phases occurs. Carbon dioxide enrichment of the steam due to bicarbonate decomposition can lead to condensate system corrosion.

Silica

Silica is a much greater problem in power boilers than in oil-field systems due to vapor phase silica deposition in turbines and process equipment. Silica can produce scaling, both directly and as a constituent of complex mineral scales on boiler tubes, and can produce sludging with resultant injection zone plugging. It is also a contributing factor to caustic embrittlement. Control of silica problems in flash systems consists primarily of maintaining silica solubility since its removal is somewhat troublesome. Silica solubility is strongly affected by alkalinity. Alkalinity should be maintained at least three times the silica content. Satisfactory operations have been maintained with silica content of up to 150 ppm.

Phosphate

Phosphate is sometimes used for alkalinity maintenance. Its presence can be beneficial as long as total hardness is well controlled.

pH

The pH level of boiler water is related both to its potential for acid corrosion, and alkalinity. Acceptable operation has been obtained at pH values from 7 to 12 with the range of 9 to 12 being optimum. Lower values indicate possible acidic corrosion while values of 13 or above indicate excess hydroxide alkalinity with both caustic corrosion and caustic embrittlement possible.

Foaming in Boilers—An Important Operational Problem

Foaming of the water in a steam boiler can be a serious nuisance to the boiler operator; it is difficult to maintain the proper water level in a boiler that is foaming badly. It can be a costly nuisance; persistent foaming causes boiler water with its dissolved and suspended solids to be entrained in the steam, which renders the steam unfit for many purposes and may cause pipes and traps to be clogged by deposition of solids—not to mention damage to turbine blades by simlar deposition. It is sometimes a dangerous nuisance. In severe cases of foaming, the steam generating zone may become filled with foam instead of water, which can lead to failure of boiler tubes due to overheating.

In a normally operating boiler, steam bubbles form on the walls of the hot tubes and rise quickly to the surface of the water where they burst. The steam passes out of the boiler, while the few minute droplets of water that may be flung upward by the bursting bubble either fall back to the surface or are trapped by baffles at the top of the boiler drum. In some cases, however, the bubbles do not burst immediately upon reaching the surface. Instead, they persist long enough for the bubbles which follow to reach the surface and push them up into the space above the surface. If this process continues so that the "original" bubbles are lifted a considerable distance above the surface of the water, we have foam.

Foam develops when steam bubbles are formed so rapidly that they arrive at the surface of the water before those ahead of them have had time to burst. Whether individual bubbles will survive long enough to make a foam depends on the "strength" and elasticity of the liquid film that forms the bubble wall. The properties of this film are determined by the kind and amount of foreign substances, both dissolved and dispersed, or suspended in the liquid. It has been demonstrated that pure liquids will not foam.

Extensive research carried out at Ohio State University and elsewhere has established the fact that water in a boiler can be made to foam by the addition of a large enough quantity of any one dissolved salt. The amount needed varies for different salts; and for any one salt, the amount needed decreases as the boiler pressure increases. Of the salts commonly found in operating boilers, sodium phosphate and sodium hydroxide appear to have the greatest foaming potential —each has about one and a half times the effect of an equal quantity of sodium chloride. For this reason, excessive amounts of alkalinity and phosphate are undesirable in boiler waters.

Suspended solids aggravate foaming; smaller amounts of dissolved salts will cause foaming when appreciable suspended matter is present. In the case of suspended solids, particle size is more important than chemical composition—the smaller the particles, the greater the effect. Suspended solids not only increase the tendency of the water to foam, but they "stabilize" the foam. That is, when solids are present, the bubble films are stronger so that the bubbles last longer and rise higher than when solids are absent. Excessive colloids, especially

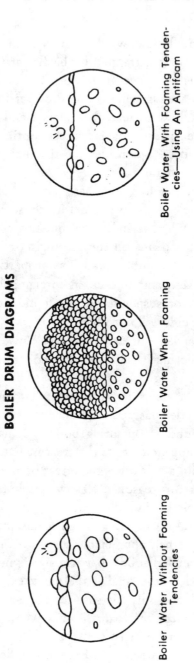

BOILER DRUM DIAGRAMS

Boiler Water With Foaming Tendencies—Using An Antifoam

Boiler Water When Foaming

Boiler Water Without Foaming Tendencies

Figure 10-20

emulsified oil, are notable foam producers; with respect to foam formation, they may be said to combine the properties of dissolved and suspended matter.

Much effort has been expended by industrial researchers and other interested groups on development and evaluation of antifoams—substances which, when added in small quantities to boiler water, prevent the formation of foam. Many different types of materials are useful for this purpose: alcohols, amids, amines, glycols, silicones and vegetable oils have been found to be effective. Only a few members of each class of compounds, however, have antifoam properties. Some antifoams have a limited usefulness because they are unstable under boiler operating conditions. Castor oil, for example, one of the best of the early antifoams, becomes saponified by alkaline boiler water and the resulting soap actually increases the tendency of the water to foam.

A considerable part of antifoam research has been devoted to determining the way in which these substances work. Two clues were apparent from the outset; substances which prevent foaming are generally insoluble in the liquids with which they are effective, and only small quantities (usually a few parts per million) are needed. Out of the many theories of antifoam action which have been advanced, two appear to have the most merit. Since both theories have been supported by experimental and photographic evidence, it is likely that both are correct.

(a). It has been observed that the steam bubbles released from heated surfaces are smaller in foaming liquids than in those which do not foam. Small bubbles tend to form on a wetted surface—i.e., one which has absorbed a layer or film of water. Some antifoams, when added to the water, are themselves absorbed on the generating surfaces. This reduces the "wettability" of the surfaces and favors formation of large bubbles.

(b). The presence of minute particles of dispersed but insoluble antifoam in the water, particularly at the surface, makes it inevitable that as bubbles reach the surface of the water they will come in contact with a few of these particles. A "patch" of antifoam is thus formed on the thin film of the bubble wall. At the periphery of this patch, the surface tension of the film is reduced—the film is "weakened." This causes quick rupture of the bubble wall, which is literally pulled apart by its own contractive forces.

Whether the antifoam action is caused by either or both of these mechanisms, the effect of introducing a good antifoam into a foaming boiler is so positive and so sudden that it can be called dramatic. Almost equally impressive is the small amount of material needed to produce so large an effect.

Boiler Water Control

In most cases, foaming can be avoided by proper control of boiler water alkalinity, total dissolved solids and suspended solids. It can be stopped, or positively prevented by use of modern antifoams. With two such effective weapons available for the boiler engineer's use, there is really no need for anyone to put up with inconvenience or expense that is caused by a foaming boiler.

11

Nonchemical Devices

Over the years, the authors have accumulated a file on nonchemical water treatment devices (NCDs) which are widely sold, are quite expensive, and are advertised with very persuasive descriptive literature. These units all have in common: no chemicals required; use little or no power; are all self-contained; and require only to be installed in a pipeline. Also in common, they have little or no effect on the water which is to be conditioned. It is claimed, however, that they will work on any kind of water equally well.

For years, suits have been brought against the makers of these products by the government and others, charging fraud, but with little success. The problem lies in the difficulty of proving a negative —how, for instance, can it be proven that there are no flying saucers? Lack of sighting is not proof positive. In addition, the claims for these devices are so broad that if the smallest detail can be substantiated, the whole claim is promoted as valid. Many critical papers have been presented, but have been ignored by the promoters, who blithely continue to market these devices, on a large scale.

A recent study report[1] was presented at the 48th annual meeting of the International Water Conference at Pittsburgh, PA in November 1987. This is a detailed study on NCDs, and is reproduced herein, with permission.

---

[1] "Field Study: Electromagnetic Device Used to Treat a Watertube Boiler," by Gary Caplan and Fred Stegmayer of Bird Archer, Inc., Coburg, Ontario, Canada.

INTRODUCTION

Nonchemical devices (NCDs) can be traced back to the late 1800s (Ref. 1). During the last century, suppliers, users and consultants have scientifically monitored, evaluated and hypothesized on the performance of these units in industrial water treatment applications. (Ref. 2, 3).

This presentation will overview NCD philosophies to ensure subject familiarity and discuss a field study where a typical unit was installed on the feedwater line to a watertube boiler.

NCD PHILOSOPHY

Why NCDs?: A review of the literature reveals that there is a plethora of NCD manufacturers. The list in Table 11-1(a) summarizes the various trade names of water treatment devices on which the authors have collected literature over the years. Table 11-1(b) includes additional trade names that have been reported in the literature. While Table 11-1 is not intended to be a complete historical record of all manufactured devices, they serve as an indicator that this technology is readily available and has been around for a long, long time.

After a period of relative obscurity, the NCD has again had a resurgence in many areas. This is probably because the conventional methods of treating boiler feedwater appear to be too complicated, expensive and/or time-consuming. Available treatment methods for *scale* prevention in steam boilers are summarized in Table 11-2.

The following representative statements have been extracted from NCD marketing literature:

- minimum or no power input
- little or no technical control required
- removal of existing scale
- reduced fuel costs (related to above point)

Table 11-1. Trade Names of Various Nonchemical Devices

(a) Linear Kinetic Cell

Dynamic Water Conditioner

Aqua cells

Aqua-corr Water Conditioners

Superior Water Conditioners

Polar Water Conditioners

Cal-C-M Dispenser

Scalemaster

Care Free Conditioners

Turbomag

The Ion Stick

Descal-A-Matic

The Purifier

Crustex

Ashbrook Water Stabilizer

EIBL Water Conditioner

Sentry Electro-Magnetic Treatment

The Stabilizer

Sullectron Electronic Water Treater

Super Ion Water Conditioner

Electrostatic Water Treater

Electrotreat

Rotomag Electromagnetic Water Conditioner

Aqua Magnetics Power Unit

Progressive Electronic Water Conditioner

The Butler Electronoltic De-scaler

(b) Colloid-o-Tron

Aquastat

Evis

Packard

Cepi

Aqua-Flo

Bon Aqua

Aquatron

Kemtune Superior

Aqua Scrubber

Aquatrol

Aquatronic

Corrotrol

Ultrastat

- reduced corrosion of metallic piping
- extended life and usefulness of equipment
- use of acid or other chemicals for scale control eliminated or reduced
- no pollution (relates to above point)
- reduction of time required to conduct chemical tests
- fast payback

With these reported benefits, it is little wonder that NCD suppliers are able to gain the attention of prospective users!

Table 11-2. Methods for Preventing Scale Formation (Ref. 4)

A. External Treatment	• chemical precipitation softening
	• ion exchange
	• membrane technology
	• **non-chemical devices**
B. Internal Chemical Treatment	• phosphate
	• chelant
	• phosphonate
	• dispersants and crystal modifiers

Note: A + B or A or B alone have been used.

Types of NCD: As with any technology, the scientific community seems compelled to classify every item or topic into types, classes and sub-classes. So it is natural to expect that NCDs too have been classified. As previous reported (Ref. 5), they are:

- catalytic (CA)
- electrostatic (ES)
- magnetic (MD)

Within each of these classes, there are also sub-classes. For example, there are MD 1, MD 2, MD 3 and MD 4 (Ref. 2) where an

MD 1   consists of a metallic housing through which water to be treated flows. In the housing is a propeller which is caused to rotate by the flowing water. An electromagnet within the housing is powered by an electronic control unit which operates on AC power and converts the power to DC.

MD 2   consists of a high strength permanent magnet which is clamped around the pipe in which water to be treated is flowing.

MD 3   consists of a shell of copper, inside which are permanent magnets. Water to be treated flows through the shell through the magnetic fields produced by the magnets.

MD 4   similar to MD 1 but contains a magnetic rotor which is caused to rotate by the flowing water.

Similarly, there are sub-classes for CA and ES units. Since this paper describes a field study dealing with an MD unit, only this technology will be described.

MAGNETIC PROPERTIES OF SUBSTANCES

In the mid-1800s, Michael Faraday revealed in his classical experiments that conducting fluids possess certain magnetic properties.

More recently, contemporary investigators have applied these fundamental concepts of magnetism and those of magnetohydrodynamics (Ref. 6) to water and MD conditioners.

Simplistically, magnetic substances* are either diamagnetic** or paramagnetic*** in nature. A number of these substances (Ref. 7) are contained in either soluble or insoluble (colloidal) form in the water circuit of a boiler.

* ferromagnetic and antiferromagnetic substances complete the classification

** diamagnetic substances have no net magnetic moment+ and are slightly repelled by an applied magnetic field

*** paramagnetic substances are attracted by an applied magnetic field and have a net magnetic moment+

+ the magnitude of the magnetic moment of an atom relates directly to the number of electrons with unpaired spins (Ref. 8)

Assuming typical feedwater quality, Table 11-3 presents the major inorganic ions in water.

Table 11-3. Magnetic Properties of Inorganic Compounds

	CO_3^{-2}	Cl^-	SO_4^{-2}	oxides
Ca^{+2}	–	–	–	–
Mg^{+2}	–	–	–	–
Na^{+1}	–	–	–	–
K^{+1}	–	–	–	+
Mn^{+2}	+	+	+	+
Fe^{+2}	+	+	+	+
Cu^{+2}	n/a	+	+	+

where – denotes diamagnetic substances
+ denotes paramagnetic substances

As noted above, the majority of these impurities are diamagnetic. Scale-forming salts with inverse solubility, like calcium carbonate and calcium sulfate, are reported to be easily treated by dislocating the electrons from their normal position in the lattice structure. Carbonate scale is then converted from a hard deposit to a soft suspension (mud) which can be removed by blowdown.

The most significant paramagnetic substances in Table 11-3 are the manganese salts plus the oxides of iron and copper. These substances are reported to be troublesome and for some waters they must be removed prior to the device (Ref. 9).

The methods for translating academic magnetic theory into the domain of industrial water treatment are proprietary and only superficially explained to the marketplace (Ref. 9). Nevertheless, NCD suppliers are given the opportunity to evaluate the water quality for the stream being conditioned and to comment on the appropriateness of the application.

ELECTROMAGNETS VS PERMANENT MAGNETS

The major difference between an electromagnet and a permanent magnet is the retentivity of the cores. Permanent magnets are initially magnetized by placing them in a coil through which a current is passed. The cores are made of retentive (magnetically "hard") materials which retain their magnetic properties for a long period of time. Electromagnets are devices in which the magnetism can be turned on or off. Therefore, the core is a nonretentive (magnetically "soft") material which maintains its magnetic properties only while current flows. With many electromagnetic devices, the intensity of the magnetic field (in gauss units) is proportional to water flow (rpms of the device impeller).

No flow, no electromagnetic field!

A changing magnetic field always produces an electric field and, conversely, a changing electric field produces a magnetic field (Ref.

10). Therefore, the many papers that discuss the application of permanent magnetic devices (Refs. 5, 6, 9–17) are applicable to electromagnetic devices.

The (electro)magnetic field when applied to a flowing conductive boiler feedwater neither alters the chemical composition of the substances contained therein, nor is it supposed to introduce any stray electric current into the water.

FIELD STUDY OF ELECTROMAGNETIC DEVICE (EMD)

This section details a field study performed on the use, for boiler feedwater conditioning, of an electromagnetic device in place of conventional water treatment. The plant trial was conducted in Ontario, Canada. The boilers produce approximately 1,000,000 lb of steam/day which is used for production and approximately 10 percent is recovered as condensate. The steam is being generated by two package-type watertube boilers, unit #2 rated at 100,000 lb per hour, and unit #1 at 45,000 lb per hour. The normal system operating pressure is 130 psi. Generally, the larger boiler is used to carry the full steam load, while the smaller boiler is operated only during low production periods and on weekends during the winter months.

Prior to the installation of the electromagnetic device (EMD), sodium zeolite softened water was used as system makeup. Based on composite water samples, the average raw water quality is shown in Table 11-4.

The soft water which, on average, contained less than 1 ppm total hardness, was fed to a holding tank which also functioned as the main condensate receiver. The combined blend of softened makeup and condensate was deaerated via a spray-type deaerator before the feedwater was introduced into the boilers.

The internal chemical treatment program originally consisted of phosphate and amine injected directly into the boiler steam drum;

Table 11-4. Average Raw Water Quality

T.D.S.	372	ppm
Conductivity	544	$u$mhos
Suspended Solids	< 0.1	ppm
Total Hardness	244	ppm as $CaCO_3$
Ca Hardness	166	ppm as $CaCO_3$
Mg Hardness	78	ppm as $CaCO_3$
SiO_2	7	ppm
Na	19	ppm
'M' Alkalinity	185	ppm as $CaCO_3$
Cl	33	ppm
SO_4	45	ppm
Fe Dissolved	0.03	ppm
Fe Suspended	< 0.1	ppm
Cu Dissolved	0.01	ppm
Cu Suspended	< 0.01	ppm
K	0.1	ppm
Mn	< 0.01	ppm
pH	8.0	

and sulfite, dispersant and antifoam fed to the storage section of the deaerator. In the Spring of 1984, the phosphate was discontinued and replaced by a blended product containing polymer, chelant and dispersant.

During the Summer of 1986, the management responsible for the operation of the boiler plant decided to install an electromagnetic device as an alternative and potentially more cost-effective approach to water treatment. In addition, it was thought that capital dollars could be saved by relocating existing NaZ softeners to the production area rather than purchasing another train. Although the EMD supplier could not provide any reference of effective performance of this device in similar type plants or of similar operating conditions in Canada or the United States, it was felt that, in view of the potential benefits to be gained, a field trial was justified. To accomplish this and to develop factual reference data, the boilers were opened during the last week of July 1986 for a detailed boiler inspection. This inspection consisted of taking photographs of the internal condition of the boilers (both mud and steam drums), as well as using fiber optics equipment for detailing internal tube surface conditions. In addition, deposit samples were obtained from various sections of the boilers for laboratory analysis to determine their composition.

Once the system went back on line with the EMD installed, the assessment was based on monitoring the following parameters:

1. Collection of composite water samples for analysis of raw water, electromagnetic device inlet/outlet, boiler water, and bottom blowdown.

2. Laboratory analysis of deposit samples obtained during regular boiler inspections.

3. Use of photographic equipment, in conjunction with fiber optics, to assess any observable changes in the internal cleanliness of the boiler.

Installation Location and Recommendations for the Use of the Electromagnetic Device

The manufacturer of the device recommended that the EMD of appropriate size be installed downstream of the deaerator, as shown in Figure 11-1. The unit itself had the following dimensions—53" x 16" x 24". The use of both softeners was discontinued. All feedwater was processed through the device. Internal chemical treatment was suspended, with the exception of the sulfite feed to the deaerator.

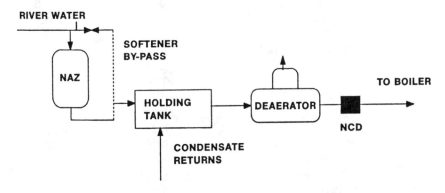

Figure 11-1. Preboiler System Schematic

Composite Water Analyses

A series of composite water samples was collected from various sample points of the steam generating system during the months of August, September and October. However, for simplicity and clarity average data only is presented in Tables 11-4, 11-5 and 11-6.

Raw Water Analyses

Statistically, throughout the study period there were no deviations from the previously reported raw water analyses (Tabel 11-4).

This data showed that rainfall and runoff did not contribute to significant variances in the quality of the river water supply source.

Table 11-5. Average Electromagnetic Device Outlet vs. Inlet Water Quality

	Average EMD Outlet	Average EMD Inlet
T.D.S.	331	334
Conductivity, mmhos	517	531
Suspended Solids	0.17	4.3
Total Hardness as $CaCO_3$	203	209.5
Ca Hardness as $CaCO_3$	141	145
Mg Hardness as $CaCO_3$	62	64.5
SiO_2	5.18	5.2
Na	27.1	30.1
'M' Alkalinity as $CaCO_3$	160	170.6
Cl	29.63	31
SO_4	51.1	57
pH	8.05	8.0
Fe Dissolved	0.028	0.2
Fe Suspended	0.125	1.4
Cu Dissolved	0.04	0.03
Cu Suspended	< 0.1	0.1
Acidified Hardness as $CaCO_3$	208.3	220.1

* All results except conductivity and pH are reported as ppm.

Relocation of Electromagnetic Device

During the first week in September, both the deaerator and boiler were opened for internal inspection. While prior to the use of the EMD the deaerator was totally free of scale, it was found that, within one month of operation, a heavy buildup of scale had occurred throughout the deaerator, with deposition thickness varying from

1/8 inch to 1 inch. The boiler also contained a heavy buildup of scale, particularly in the mud drum (more than one wheelbarrowful of loose scale was removed). To retard any further buildup of scale in the deaerator, the EMD was relocated upstream of the deaerator and the steam pressure to the deaeratoring dome was reduced.

Table 11-6. Averaged Composite Analyses*: Boiler Water and Blowdown Tank

	Boiler #1	Boiler #2	Blowdown Tank
T.D.S.	-	1505	1437
Conductivity, mmho	-	2200	2112.5
Suspended Solids	166	383.3	665
Total Hardness as $CaCO_3$	286	245.2	266.6
Ca as $CaCO_3$	202	188	232
Mg as $CaCO_3$	84	57.2	34.6
SiO_2	-	2.64	1.97
'P' as $CaCO_3$	-	94	67.8
'M' as $CaCO_3$	-	223	152
Cl	357	329.7	250
SO_4	582	440.2	405.2
PO_4	< 0.1	< 0.1	< 0.1
pH	-	9.8	9.7
Fe Dissolved	-	0.09	0.03
Fe Suspended	22.2	9.9	19.1
Cu Dissolved	-	0.05	0.02
Cu Suspended	< 0.1	0.01	225
Acidified Hardness as $CaCO_3$	406	1171.6	1254.4

* All results except conductivity & pH are reported as ppm.

Boiler Inspection After Three Months of Operation

The next detailed boiler inspection was performed on October 31, 1986, approximately three months after the original installation of the EMD. This inspection showed heavy buildup of whitish deposit, particularly in the mud drum. Most of the scale was approximately 1/8 inch thick and was primarily calcium carbonate (with some iron). Large chunks, up to 1 inch thick, had also broken loose. Some tubes, especially horizontal and inclined runs, had partial blockages with hard, chunky scale. The fiber optics inspection revealed a significant increase of deposit buildup on the tube surfaces to the extent that some were half plugged with scale.

Tube Failure

In spite of the progressive buildup of deposits, the plant's decision was to carry on with the EMD until the next major inspection scheduled during the Christmas season. Unfortunately, the internal conditions continued to deteriorate to the point where, in mid-December, a boiler failure occurred, necessitating the complete shutdown of the #2 boiler.

DISCUSSION

Water Quality

The installed EMD was of the MD 1 type as referred to earlier. In analyzing the EMD inlet water quality (Table 11-5), and using the criteria reported by Carpenter (Ref. 9) that 200 ppm of total dissolved solids must be present for each ppm of iron and manganese, we are able to confirm in fact that the evaluation was conducted on "acceptable" water quality. Upon close examination of the inlet and outlet streams, and using analysis of variance (ANOVA) (Ref. 18), the only significant element concentration that changed was the dissolved iron. In our study, the dissolved iron was consistently lower in the outlet than the EMD inlet. This observation is contrary to other reports where the opposite phenomenon was occurring (Ref. 3, 11).

Sludge Recovery Tests

Sludge recovery testing is one method of performing a material balance of the boiler water. This simplistic technique has been frequently referenced (19-22) because it offers plant personnel an easy way to determine the rate of transport of precipitated solids within the boiler. If boiler tubes are to remain clean, "whatever solids go in—must come out." The equation used is as follows:

$$\% \text{ sludge recovery} = \frac{\text{Impurity from bottom blowdown} \times 100}{\text{Impurity from feedwater} \times \text{COC}} \quad (1)$$

Using the data in Tables 11-5 and 11-6, and selecting the most soluble salts like chloride and sulfate as the monitored impurity, the cycles of concentration (COC) can be calculated as follows:

	Feedwater*	Blowdown Tank	COC
Chloride	29.63 ppm	250 ppm	8.4
Sulfate	51.50 ppm	405 ppm	7.9
*EMD outlet water quality		Average	8.2
(Table 11-5)			

Substituting the value of 8.2 for the COC in equation (1) and monitoring sludge recovery on total acidified hardness constituents, the following equation is developed:

$$\% \text{ sludge recovery} = \frac{1254.4 \times 100}{208.3 \times 8.2} = 73.4\% \quad (2)$$

Although this method is crude, there is no doubt this value is meaningful. The conclusion that can therefore be drawn from the water analysis is that hardness salts of calcium and magnesium are not being removed from the boiler. In other words, 100 percent sludge recovery based on total hardness should yield a blowdown tank value of 1708 ppm. The difference in total hardness solids of 454 lb/mm lb of steam must be accumulating within the boiler!

Deposit Analysis

Table 11-7 identifies the composition of deposits removed from boiler #2 over 4.5 years. While the boiler was on a phosphate-based program, the major components were as expected—hydroxyapatite and serpentine. No calcium carbonate was present. Although the treatment program changed in mid-1984, residual deposits of the same composition were still present during the October 1985 inspection. Their composition was confirmed by X-ray diffraction analysis. The November 1986 sample removed during the October inspection showed a dramatic change in the nature of the deposit.

Again, X-ray analysis confirmed that the polymorphs of calcium carbonate and calcium sulfate were calcite (and not aragonite as has been reported in another study [Ref. 16]), and anhydrite, respectively. The customer was still happy with the "performance" of the unit.

At the final inspection in December 1986, it was found that at least 400 tubes in boiler #2 required replacement, which necessitated an unplanned shutdown and severely reduced plant output! The January 1987 data is typical of the deposit analysis from several failed tube sections that were submitted for analysis and photographic documentation.

Photographic Evidence

Photograph No. 1 shows a thin black ring of deposition directly adjacent to the inside tube surface (denoted as I.W.), with a considerably thicker portion of greyish-white deposit above it (denoted as O.W.). X-ray analysis revealed that the blackened deposit was formed during the phosphate treatment regime as the primary components were hydroxyapatite, serpentine, and some calcite. The outer white layer, which formed after the installation of the EMD, is over 80 percent calcite and 5–10 percent brucite ($Mg(OH)_2$). The X-ray diffraction traces are shown in Figure 11-2.

Table 11-7. Deposit Analysis—Boiler #2

	July '82	Aug. '84	Oct. '85	Nov. '86	January, 1987 I.W.°	O.W.°°
$CaCO_3$	0.0	0.0	0.0	52.7	6.8	81.8
$CaSO_4$	0.0	0.0	0.0	24.3	13.3	4.2
$Ca_3(PO_4)_2Ca(OH)_2$†	48.9	47.6	47.9	0.9	27.2	0.0
$CaSiO_3$	1.35	8.1	0.4	5.2	0.0	0.0
$MgOSiO_2{}^2H_2O$‡	21.3	17.3	29.9	0.0	22.0	3.2
$Mg(OH)_2$	3.5	0.0	1.7	9.7	3.5	2.4
Fe_2O_3	11.4	10.4	8.9	0.3	4.2	0.4
CuO						

*   Based on Wet Chemical Analysis

†   hydroxyapatite

‡   serpentine

°   inner tube surface

°°   outer tube surface

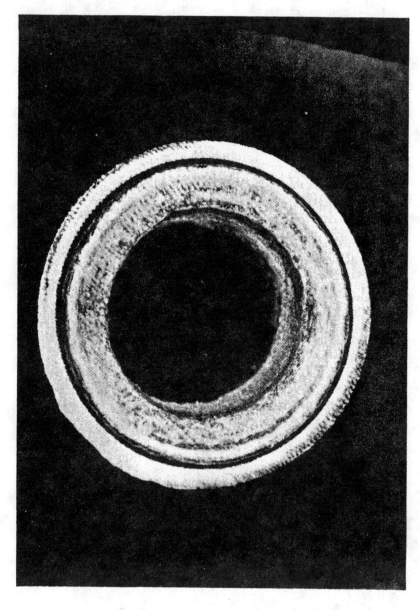

Photograph No. 1.   From Sectional View of Failed Tube Depicting Two Distinct Layers of Scale

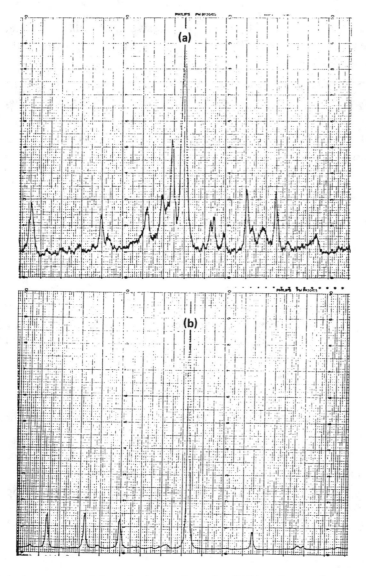

(a) Deposit was removed from the inner tube surface (I.W.); major constituents were hydroxyatite and serpentine.

(b) The deposit from the outside tube surface (O.W.) was primarily calcite.

Figure 11-2. X-Ray Diffraction Patterns of January 1987 Deposits

Photographs 2 to 6 show the progressive accumulation of deposit in areas of low circulation, leading to increasingly restricted flow and, eventually, complete blockage of the tube.

It is important to note that, even after this overwhelming evidence had been presented, the manufacturer of the EMD insisted that the failure was not due to lack of performance of the EMD. Rather, they stated that it was working too effectively in removing old deposit and, consequently, leading to the blockage of "some" tubes.

This is a favorite argument put forth by many NCD suppliers. However, three facts are presented which refute this claim.

(a)   If the sludge recovery results were as high as 100 percent or higher, maybe this situation was happening. However, at 73 percent sludge recovery, deposition was occurring in the boiler.

(b)   The flue gas temperature for boiler #2 under clean conditions ran at 468°F whereas, in November 1986, at similar load, it was 501°F. This indicates fouling, not stripping. Plant records also confirm an increase in fuel costs!

(c)   Ultimately, over 600 tubes required replacement!

For the majority of the boiler water samples collected, the boilers were operating at increased blowdown since the neutralized conductivity was 2200 mmho (1700-2600). When under a full chemical treatment program, the specified limits were higher at 2500-3000 mmho. Based on items (a) and (b) even these more favorable operating conditions between August and December didn't clean up the boilers.

Conclusion

At last count, 640 tubes have been replaced. The direct cost for retubing the boiler has exceeded $100,000. The plant returned to a full chemical treatment program after both boilers were acid cleaned to restore heat transfer cleanliness. This particular EMD in this particular plant

- failed to replace water softening units

- failed to generate a loose washable mud of finely divided particles

- failed to remove existing scale

- failed to reduce fuel costs

- failed to extend equipment life

- failed to deliver a fast payback

This field study is more detailed than other documented investigations from non-Eastern block sources (Soviet Union in particular) (Refs. 6, 17, 23*, 24). However, most of these reports relate to cooling systems and few to watertube boilers. When similar devices fail to perform as promised, pseudo scientific explanations are always provided. "Inappropriate experimental conditions" is a routine response to negative test results.

Was this unit just a "gadget" (Refs. 1, 25-30) or was it really a Non-Credible Device?

In conclusion, we have learned that

- very few documented field studies are available in spite of the overwhelming number of technical papers and testimonials

- iron poisoning of the crystal lattice for $CaCO_3$ and $CaSO_4$ was not the mechanism for scale control

- no descaling mechanism was observed during the test period.

---

*This study was later deemed a failure (Ref. 31).

Photograph 2. Progressive Scale Accumulation (Note Scale Bridging the Tube)

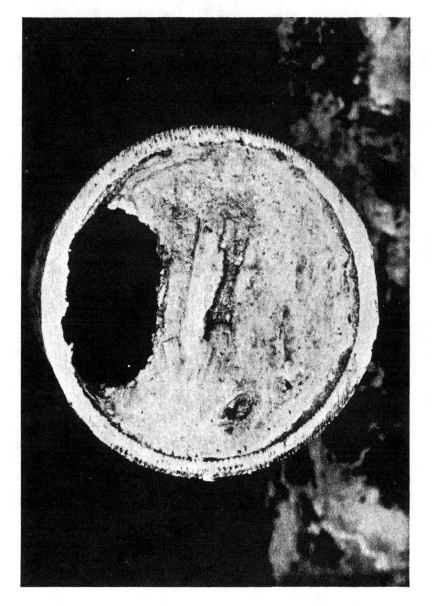

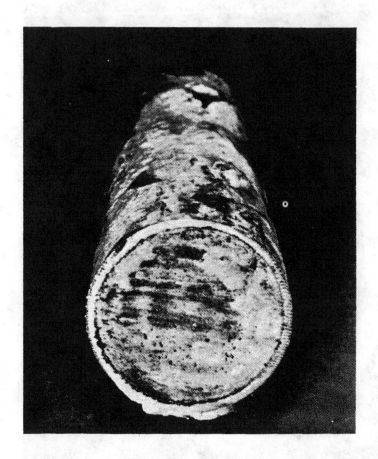

Photograph 4. Total Blockage

Photograph 5. Example of Ruptured Boiler Tube

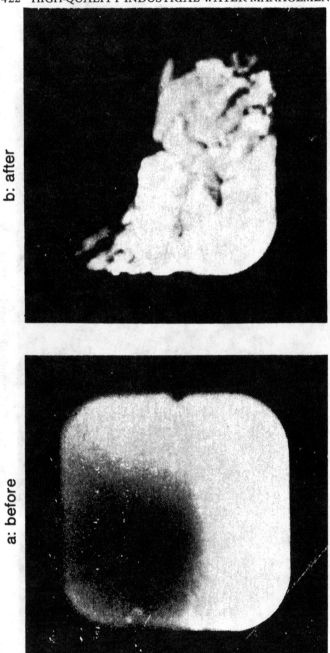

b: after

a: before

Photograph 6.   Boiler #2: Fiber Optics Photographs Taken Before EMD Trial (July '86) and After First Major Inspection (October '86)

References

1. B. Q. Welder, E. P. Partridge, "Practical Performance of Water-Conditioning Gadgets," *Industrial and Engineering Chemistry*, p. 954, Vol. 46, No. 5.
2. G. J. C. Limpert, J. L. Raber, "Tests of Non-Chemical Scale Control Devices in a Once-Through System," Corrosion/85, Paper No. 250, National Association of Corrosion Engineers, Houston, Texas, 1985.
3. P. Puckorius, "Mechanical Devices for Water Treatment: Just How Effective Are They?" *Power*, p. 60-62, January 1981.
4. Bird Archer Inc., ed. G. Caplan, *The Guide to Water Treatment*, 1987.
5. C. E. Gruber, D. D. Carda, "Performance Analysis of Permanent Magnet Type Water Treatment Devices," Final Report issued to the Water Quality Association, South Dakota School of Mines and Technology, July 1981.
6. E. Raisen, "The Control of Scale and Corrosion in Water Systems Using Magnetic Fields," Corrosion/84, Paper No. 117, National Association of Corrosion Engineers, Houston, Texas, 1984.
7. *CRC Handbook of Chemistry and Physics*, 48th ed., p. E107-E112.
8. E. Cartmell, G. W. A. Fowles, *Valency and Molecular Structure*, 3rd ed., p. 224, 1966.
9. R. K. Carpenter, "Magnetic Treatment Does Work," Corrosion/85, Paper No. 252, National Association of Corrosion Engineers, Houston, Texas, 1985.
10. *McGraw Hill Encyclopedia of Science and Technology*, Vol. 4, p. 552, 1977.
11. K. W. Busch, M. A. Busch, D. H. Parker, R. E. Darling, J. L. McAtee Jr., "Studies of a Water Treatment Device That Uses Magnetic Fields," Corrosion/85, Paper No. 251, National Association of Corrosion Engineers, Houston, Texas, 1985.
12. D. Hasson, D. Bramson, "The Performance of a Magnetic Water Conditioner Under Accelerated Scaling Conditions," Progress of Fouling Conference. University of Nottingham, 1981.
13. T. Vermeiren, "Magnetic Treatment of Liquids for Scale and Corrosion Prevention," *Corrosion Technology*, p. 214-219, July 1958.
14. R. M. Clyburn, "Non-Chemical Water Treating: Comparing the Theories," Research Report CIVE 6398, University of Houston, Texas, January 26, 1983.
15. Evaluation of the Principles of Magnetic Water Treatment, American Petroleum Institute, API Publication 960, September, 1985.
16. K. S. Narasaih, "Magnetic Treatment of Water—A Solution to Prevent Corrosion?" *Water and Pollution Control*, p. 34-37, June 1970.

17.   S. G. Hibben, Magnetic Treatment of Water, Advanced Research Projects Agency of the Department of Defence ARPA Order No. 1623-3, January 30, 1973.
18.   R. E. Miller, "Part 4, Means and Variances," *Chemical Engineering,* January 21, 1985, p. 107-110.
19.   D. G. Cuisia, "The Use of Sulphonated Styrene Copolymers for Boiler Scale Control," International Water Conference 39th Annual Meeting, Pittsburgh, Pennsylvania, p. 289, October 1978.
20.   N. L. Fleck, "Iron Transport Chemical Reduces Deposits, Improves Boiler Efficiency," *Chemical Processing,* p. 42-43, March 1982.
21.   J. A. Gray, "Trends in Boiler Water Treatment," *Modern Power and Engineering,* December 1969, January 1970.
22.   J. A. Gray, "Energy Conservation Through Effective Water Treatment," International District Heating Association, New Hampshire, 1984.
23.   J. F. Grutsch, J. W. McClintock, "Corrosion and Deposit Control in Alkaline Cooling Water Using Magnetic Water Treatment at Amoco's Largest Refinery," Corrosion/84, Paper 330, National Association of Corrosion Engineers, Houston, Texas, 1984.
24.   J. F. Wilkes, "Water Conditioning Devices—An Update," International Water Conference 40th Annual Meeting, Pittsburgh, Pennsylvania, p. 161-167. October 1979.
25.   J. C. Dromgoole, "The Fatal Lure of Water Treating Gadgets," International Water Conference 40th Annual Meeting, Pittsburgh, Pennsylvania, p. 169-173. October 1979.
26.   R. M. Westcott, "Nonchemical Water Treating Devices," *Materials Performance,* p. 40-42, November 1980.
27.   Anon (editorial), "Why be a Gadget Sucker?" *Corrosion,* Vol. 16, No. 7, p. 7, 1960.
28.   J. C. Dromgoole, M. C. Forbes, "A New Answer to The Water Treating 'Gadget' Problem," Winter Meeting, Cooling Tower Institute, Houston, Texas, Paper TP 265A, 1983.
29.   M. G. Fontana, N. D. Greene, *Corrosion Engineering,* McGraw Hill, p. 197.
30.   Dearborn Chemical Problem Solver No. 28, Problem: Gadgets for Water Treatment.
31.   Cooling Tower Institute, 1986 Winter Meeting, Committee Minutes.

Appendix

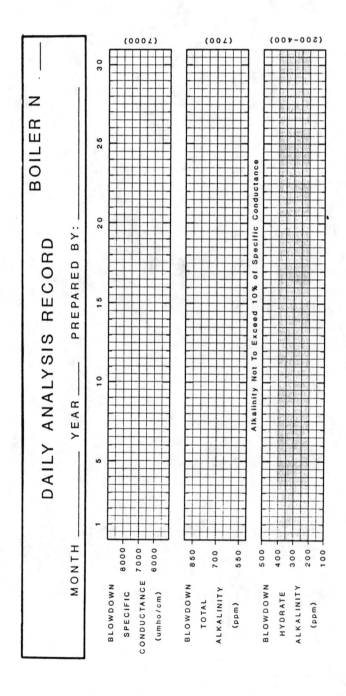

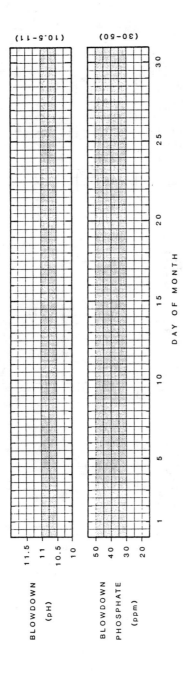

Daily Analysis Record Sample Form

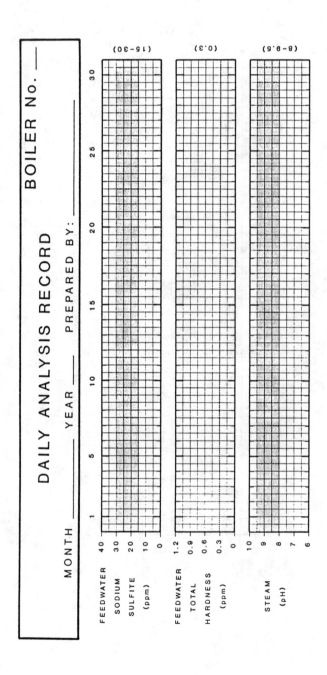

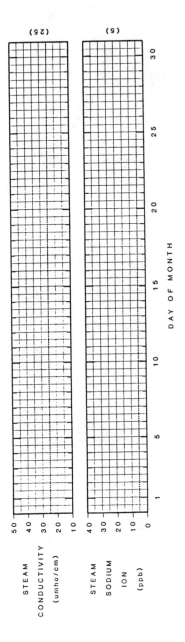

Daily Analysis Record Sample Form

BOILER PLANT MONTHLY REPORT

MONTH _____ DAY _____ YEAR _____

PREPARED BY: _____

AVERAGE MAKEUP ANALYSIS (DAILY)					
SILICA (ppm)					
TOTAL ALKALINITY (ppm)					
TOTAL INFLUENT HARD (ppm)					
SOFTENER REGENERATION					
SALT USED (lbs.)					
TOTAL EFFLUENT HARD (ppm)					

NOTES

CHEMICAL USE	BOILER #1	BOILER #2	BOILER #3
#1 CHEMICAL (lbs.)			
#2 CHEMICAL (lbs.)			
#3 CHEMICAL (lbs.)			
STEAM PRODUCED (lbs.)			
HOURS OF OPERATION			

PLANT TOTALS

STEAM (lbs.)	
FEEDWATER (lbs.)	
MAKEUP WATER (gals.)	
CONDENSATE (gals.)	
AMINES (lbs.)	
SODIUM SULFITE (lbs.)	
CHEMICAL #1	
CHEMICAL #2	
CHEMICAL #3	
TOTAL FUEL (gals or millions cu ft gas)	

Comments: _____

AVERAGE CORROSION ANALYSIS (WEEKLY)		NOTES
Cu IN CONDENSATE (ppm)		(0.050 ppm max.)
Fe IN CONDENSATE (ppm)		(0.100 ppm max.)
AMINE RESIDUAL (ppm)		

PERCENTAGES	THIS MONTH	LAST MONTH	DIFF
FUEL (per million lbs. steam)			
% MAKEUP			
% CONDENSATE RETURN			
% BLOWDOWN			
SALT (lbs. per million gals water)			
SCAVENGER (lbs. per million gals water)			
AMINE (lbs. per million lbs steam)			
CHEMICAL (lbs. per million lbs. steam)			

Boiler Plant Monthly Report Sample Form

Molecular and Equivalent Weights of Ions Commonly Found in Water

To assist the engineer in preparation of process water or wastewater treatment installations, some basic data that can be used to establish design parameters is presented.

Compounds	Formula	Molecular Weight	Equivalent Weight	Substance CaCO₃ Equivalent	CaCO₃ Equivalent to Substance
				Multiply By	
Aluminum Sulfate (anhydrous)	$Al_2(SO_4)_3$	342.1	57.0	0.88	1.14
Aluminum Hydrate	$Al(OH)_3$	78.0	26.0	1.92	0.52
Aluminum Oxide (Alumina)	Al_2O_3	101.9	17.0	2.94	0.34
Sodium Aluminate	$Na_2Al_2O_4$	163.9	27.3	1.83	0.55
Barium Sulfate	$BaSO_4$	233.4	116.7	0.43	2.33
Calcium Bicarbonate	$Ca(HCO_3)_2$	162.1	81.1	0.62	1.62
Calcium Carbonate	$CaCO_3$	100.1	50.0	1.00	1.00
Calcium Chloride	$CaCl_2$	111.0	55.5	0.90	1.11
Calcium Hydrate	$Ca(OH)_2$	741	37.1	1.35	0.74
Calcium Oxide	CaO	56.1	28.0	1.79	0.56
Calcium Sulfate (anhydrous)	$CaSO_4$	136.1	68.1	0.74	1.36
Calcium Sulfate (gypsum)	$CaSO_4 \cdot 2H_2O$	172.2	86.1	0.58	1.72
Calcium Phosphate	$Ca3(PO4)_2$	310.3	51.7	0.97	1.03
Ferrous Sulfate (anhydrous)	$FeSO_4$	151.9	76.0	0.66	1.52
Ferric Sulfate	$Fe_2(SO_4)_3$	399.9	66.7	0.75	1.33
Magnesium Oxide	MgO	40.3	20.2	2.48	0.40
Magnesium Bicarbonate	$Mg(HCO_3)_2$	146.3	73.2	0.68	1.46
Magnesium Carbonate	$MgCO_3$	84.3	42.2	1.19	0.84
Magnesium Chloride	$MgCl_2$	95.2	47.6	1.05	0.95

Name	Formula				
Magnesium Hydrate	Mg(OH)$_2$	58.3	29.2	1.71	0.58
Magnesium Phosphate	Mg$_3$(PO$_4$)$_2$	262.9	43.8	1.14	0.88
Magnesium Sulfate (anhydrous)	MgSO$_4$	120.4	60.2	0.83	1.20
Magnesium Sulfate (Epsom Salts)	MgSO$_4$ • 7H$_2$O	246.5	123.3	0.41	2.47
Manganese Chloride	MnCl$_2$	125.8	62.9	0.80	1.26
Manganese Hydrate	Mn(OH)$_2$	89.0	44.4	1.13	0.89
Potassium Iodide	KI	166.0	166.0	0.30	3.32
Silver Chloride	AgCl	143.3	143.3	0.35	2.87
Silver Nitrate	AgNO$_3$	169.9	169.9	0.29	3.40
Silica	SiO$_2$	60.1	30.0	1.67	0.60
Sodium Bicarbonate	NaHCO$_3$	84.0	84.0	0.60	1.68
Sodium Carbonate	Na$_2$CO$_3$	106.0	53.0	0.94	1.06
Sodium Chloride	NaCl	58.5	58.5	0.85	1.17
Sodium Hydrate	NaOH	40.0	40.0	1.25	0.80
Sodium Nitrate	NaNO$_3$	85.0	85.0	0.59	1.70
Tri-sodium Phosphate	Na$_3$PO$_4$ • 12H$_2$O	380.2	126.7	0.40	2.53
Tri-sodium Phosphate (anhydrous)	Na$_3$PO$_4$	164.0	54.7	0.91	1.09
Disodium Phosphate	Na$_2$HPO$_4$ • 12H$_2$O	358.2	119.4	0.42	2.39
Disodium Phosphate (anhydrous)	Na$_2$HPO$_4$	142.0	47.3	1.06	0.95
Monosodium Phosphate	NaH$_2$PO$_4$ • H$_2$O	138.1	46.0	1.09	0.92
Monosodium Phosphate (anhydrous)	NaH$_2$PO$_4$	120.0	40.0	1.25	0.80
Sodium Metaphosphate	NaPO$_3$	102.0	34.0	1.47	0.68
Sodium Sulfate	Na$_2$SO$_4$	142.1	71.0	0.70	1.42
Sodium Sulfite	Na$_2$SO$_3$	126.1	63.0	0.79	1.26

Glossary

Certain license is taken with strict semantic interpretation of the following words, to present definitions related specifically to water treatment technology.

ABSORPTION, ABSORBENT: To take up in the fashion of a sponge, in the pores of a solid body.

ACID: A solution that contains hydrogen ions and has a pH less than 7.

ACID ATTACK: Corrosion caused by an acid.

ACIDITY: The quantitative capacity of aqueous media to react with hydroxyl ions.

ACTIVATED CHARCOAL: Charcoal that has been processed to increase its surface area; used as a filter medium.

ADSORBENT: A synthetic resin possessing the ability to attract and to hold charged particles.

ADSORPTION: The adhesion of a thin layer of molecules to surfaces of solid bodies or liquids.

ALKALI: A solution that contains hydroxide ions and has a pH more than 7.

ALKALINITY: An expression of the total basic anions (hydroxyl groups) present in a solution. It also represents, particularly in water analysis, the bicarbonate, carbonate, and occasionally, the borate, silicate, and phosphate salts which will react with water to produce the hydroxyl groups.

ALLOY: A substance composed of two or more metals combined by heating.

AMINE: A compound derived from ammonia by replacement of hydrogen with one or more hydrocarbon radicals. Commonly used for protection of condensate systems.

ANION: A negatively charged ion.

ANION EXCHANGE: Exchange of anions for hydroxyl ions.

ANION INTERCHANGE: Displacement of one negatively charged particle by another on an anion exchange material.

ANION RESIN BEADS: Resin beads in an ion exchanger that attract negative ions.

ANODE: The positive terminal of an electrolytic cell.

ANTHRACITE COAL: Hard coal that can be used as a filter medium.

APPROACH TEMPERATURE: In a cooling tower, the difference between the temperature of the cooled water and the wet bulb.

ATTRITION: Rubbing of one particle against another in a resin bed; frictional wear that will affect the size of resin particles.

BACKWASH: That part of the operating cycle of an ion-exchange process where a reverse upward flow of water expands the bed, effecting physical changes such as loosening the bed to counteract compacting, stirring up and washing off light insoluble contaminants to clean the bed, or separating a mixed bed into its components to prepare it for regeneration.

BASE-EXCHANGE: Property of trading cations by insoluble, naturally occurring materials (zeolites). Developed to a high degree of specificity and efficiency in synthetic resin adsorbents.

BATCH OPERATION: Use of ion exchange resins to treat a solution in a container to remove ions. Accomplished by agitation of the solution and subsequent decanting of the treated liquid. Any process operation that is not continuous.

BED: Ion exchange resin contained in a column.

BED DEPTH: Height of the resinous material in the column after the exchanger has been conditioned for effective operation.

BED EXPANSION: Effect produced during backwashing: resin particles become separated and rise in the column. Expansion of the bed due to the increase in the space between resin particles may be controlled by regulating backwash flow.

BICARBONATE: A component of alkalinity.

BICARBONATE ALKALINITY: Presence of a solution of hydroxyl ions resulting from the hydrolysis of carbonates or bicarbonates. When these salts react with water, a strong base and a weak acid are produced, and the solution is alkaline.

BIOCHEMICAL OXYGEN DEMAND (BOD): Quantity of oxygen required for biological and chemical oxidation of waterborne substances under test conditions. Refer to Chemical Oxygen Demand (COD).

BIODEGRADEABLE: A quality leading to chemical degradation of a substance to a benign form when exposed to the environment.

BIOLOGICAL DEPOSITS: Deposits composed of either microscopic organisms, such as slimes, or macroscopic organisms, such as barnacles. Slimes usually are gelatinous or filamentous.

BLOWDOWN (BLEED): Removal of a portion of water contained in a boiler drum, cooling tower basin, etc., to reduce concentration of impurities.

BOILER WATER: A term construed to mean a representative sample of the circulating boiler water, after generated steam has been separated, and before incoming feedwater or added chemical is mixed with it.

BOILING OUT: Treatment of the waterside of boilers and piping systems with hot, strong chemicals to remove oil, mill scale, and deposits.

BRACKISH WATER: Water having a dissolved matter content in the range of approximately 1,000 to 30,000 mg/liter.

BREAKTHROUGH: First appearance in a solution flowing from an ion exchange unit of unadsorbed ions similar to those that deplete the activity of the resin bed. Breakthrough is an indication that regeneration of the resin is necessary.

BRINE: Water having more than approxiamtely 30,000 mg/liter of dissolved matter.

BUFFER: A chemical which tends to stabilize the pH of a solution preventing any large change on the addition of moderate amounts of acids or alkalies.

CALCIUM CARBONATE EQUIVALENT: An expression denoting the activity of a chemical equivalent to the activity of calcium carbonate.

CAPACITY: Adsorption ability possessed by ion exchange materials. This quality may be expressed as kilograins per cubic foot, gram-milliequivalents per gram, pound-equivalents per pound, gram-milliequivalents per milliliter, etc., where numerators of the ratios represent weight of the ions adsorbed and denominators represent weight or volume of the adsorbent.

CARBONACEOUS EXCHANGERS: Ion exchange materials prepared by the sulfonation of coal, lignite, peat, etc.

CARBONATE: Component of alkalinity.

CARRYOVER: Process that transports impurities in boiler water into steam areas in water droplets.

CATALYST: A substance which by its presence accelerates a chemical reaction without itself entering into the reaction.

CATHODE: Negative terminal of an electrolytic cell.

CATION: Positively charged ion.

CATION RESIN BEADS: Resin beads in an ion exchanger that attract positive ions.

CAUSTIC: Solution that contains hydroxyl ions and has a pH greater than 7.

CAUSTIC ATTACK: Corrosion caused by an alkali.

CHANNELING: Cleavage and furrowing of a bed caused by faulty operational procedures. The solution being treated, following paths of least resistance, runs through these furrows, and fails to make adequate contact with the bed.

CHELATE (CHELANT): Chemical structure that holds a metallic ion in a ring of five or six atoms. The metallic ion is "sequestered."

CHELATION: The property of a chemical which when dissolved in water keeps the hard water salts in solution and thus prevents the formation of scale. Generally applied to organic compounds such as the salts of ethylenediaminetetraacetic acid (EDTA).

CHEMICAL OXYGEN DEMAND (COD): Amount of oxygen, expressed in parts per million, consumed under specific conditions during oxidation of waterborne organic and inorganic matter.

CHLORINE RESIDUAL: Amount of available chlorine present in water after the addition of chlorine.

CHLORINE REQUIREMENT: Amount of chlorine, expressed in parts per million, required to achieve the objectives of chlorination.

CLARIFIER: Water treatment equipment used to remove large concentrations of suspended solids from raw water.

CLOSED SYSTEM: Water system that is not exposed to atmosphere.

COLLOID: A system of which one phase is made up of particles (colloidal materials) having dimensions of 10 to 10,000 angstroms and which is dispersed in a different phase (dispersion medium). When a colloid is present in water or other liquid, it diffuses very little or not at all through a membrane.

COLOR-THROW: Discoloration of liquid passing through an ion exchange material; the flushing from resin interstices of traces of colored organic reaction intermediates.

COLUMN OPERATION: Upflow or downflow of solution through ion exchange resins.

COMBINED AVAILABLE CHLORINE RESIDUAL: Chlorine residual combined with ammonia, nitrogen, or nitrogenous compounds.

CONCENTRATION: An expression of the amount of a substance in a solution, expressed as PPM or gms/liter.

CONDUCTIVITY: Measure of electrical conductivity expressed as mhos, the inverse of resistivity.

CORROSION PRODUCTS: Compounds formed by either chemical or electrochemical reaction between a metal and its environment.

COUNTERFLOW: Applied to demineralizer systems when regenerants flow in a direction opposite to the normal flow of treated water. Applied to heat exchangers when the cooled fluid moves in a direction opposite to the coolant.

CROSSLINKAGE: Degree of bonding of a monomer, or set of monomers, to form an insoluble tri-dimensional resin matrix.

CYCLE: A complete course of ion exchange operation. For instance, a complete cycle of cation exchange would involve: exhaustion of regenerated bed, backwash, regeneration and rinse to remove excess regenerant.

CYCLES OF CONCENTRATION: A ratio of the concentration of solids in boiler, or cooling tower, water to the concentration of the same solids in influent water.

DEAERATION: Removal of dissolved air from a liquid (usually water).

DEASHING: The removal from solution of inorganic salts by means of absorption by ion exchange resins of both the cations and the anions that comprise the salts. See Deionization.

DEGASIFICATION: Removal of any dissolved gaseous product from a liquid.

DEIONIZATION: Removal of charged constituents or ionizable salts (both inorganic and organic) from solution.

DEMINERALIZER: See Ion Exchanger.

DEMINERALIZING: See Deionization.

DENSITY: (In demineralization) Weight of a given volume of exchange material, backwashed and in place in a column.

DIATOMACEOUS EARTH: Mined filter medium composed of the siliceous remains of minute planktonic unicellular or colonial algae.

DIMER: A molecule composed of two identical simpler molecules, or a polymer derived from two identical monomers.

DISPERSANT: A substance added to water to prevent the precipitation and agglomeration of solid scale. Generally a protective colloid.

DISSOCIATION: See Ionization.

DISTRIBUTION RATIO: Applied to amines, the ratio of amines in a liquid to amines in steam surrounding the liquid.

DISSOLVED MATTER: That matter, exclusive of gases, which is dispersed in water to give a single phase of homogenous liquid.

DOWNFLOW: Conventional direction of solutions processed in ion exchange column operation.

EFFICIENCY: Effectiveness of the operational performance of an ion exchanger. Efficiency in the adsorption of ions is expressed as the quantity of regenerant required to remove a specified unit weight of adsorbed material, e.g., pounds of acid per kilograin of salt removed.

EFFLUENT: Solution from an ion exchange column. The solution that exits any process.

ELECTRODIALYSIS: Process of causing water to flow through a membrane by applying an electrical potential across the membrane.

ELECTROLYTE: A chemical compound that dissociates or ionizes in water to produce a solution that will conduct an electric current; an acid, base, or salt.

ELECTRICAL CONDUCTIVITY: The reciprocal of the resistance in ohms measured between opposite faces of a centimeter cube of an aqueous solution at a specified temperature.

ELUTION: The stripping of adsorbed ions from an ion exchange material by solutions containing other ions in relatively high concentrations.

EQUILIBRIUM REACTIONS: The interaction of ionizable compounds impelling them to revert to substances from which they were formed. When a balance is reached both reactions and products are present in definite ratios.

EQUIVALENTS PER MILLION (epm): A unit chemical equivalent weight of solute per million unit weights of solution. Concentration in equivalents per million is calculated by dividing concentration in parts per million by the chemical combining weight of the substance or ion.

Note: This unit also is called "milliequivalents per liter" and "milligram equivalents per kilogram." The latter term is precise. The former will be in error if specific gravity of the solution is not exactly 1.0.

EXCHANGE VELOCITY: The rate one ion is displaced from an exchanger in favor of another.

EXHAUSTION: The state in which the resin is no longer capable of useful ion exchange; the depletion of the exchanger's supply of available ions. The exhaustion point is determined arbitrarily in terms of: (a) a value in parts per million of ions in the effluent solution; (b) the reduction in quality of the effluent water determined by a conductivity bridge which measures the electrical resistance of the water.

FILTER: A porous device that traps solids but allows a fluid to pass through.

FILTER MEDIUM: The part of a filter that traps solids.

FINES: Extremely small particles of ion exchange materials.

FLASHING: Release of water vapor (steam) when pressure on the water is less than pressure of the vapor phase.

FLOW RATE: Volume of solution passing through a specified quantity of resin within a specified time. Usually expressed in terms of gallons per minute per cubic foot of resin, as milliliters per minute per milliliter of resin, or as gallons per square foot of resin per minute.

FOAMING: Evolution of froth due to agitation and increased surface tension of boiler water when dissolved solids concentrations are high.

FOULING: (Biological) Clogging of a heat exchange surface by slime or other types of biological growth. Deposits of either airborne or waterborne particles on a heat exchange surface.

FREE AVAILABLE CHLORINE RESIDUAL: Residual consisting of hypochlorite ions, hypochlorous acid, or a combination of the two.

FREEBOARD: Space provided above a resin bed in an ion exchange column to allow for expansion during backwashing.

FREE MINERAL ACIDITY: The quantitative capacity of aqueous media to react with hy ions to pH 4.3.

FRESH WATER: Water having less than approximately 1,000 mg/liter of dissolved matter.

GEL: Applied to ion exchange resins having little true porosity.

GEAINS PER GALLON: An expression of concentration of material in solution, generally in terms of calcium carbonate. One grain (as calcium carbonate) per gallon is equivalent to 17.1 parts per million.

GREENSANDS: Naturally occurring materials primarily composed of complex silicates, that possess ion exchange properties.

HARDNESS: A characteristic of water generally accepted to represent the total concentration of calcium and magnesium ions.

Note: Originally hardness was understood to be the capacity of a water for precipitating soap. Soap is precipitated chiefly by calcium and magnesium ions commonly presented in water but may also be precipitated by ions of other polyvalent metals, such as iron, manganese and aluminum, and by hydrogen ions.

Hardness was originally measured by the amount of soap required to produce a stable lather. Measurement is usually made on a water sample, the alkalinity of which has been adjusted to eliminate the effect of hydrogen ions.

Hardness is expressed fundamentally in terms of the chemical equivalents of metal ions capable of precipitating soap. It has commonly been expressed in terms of the equivalent amount of calcium carbonate.

HARDNESS AS CALCIUM CARBONATE: The expression ascribed to the value obtained when the hardness-forming salts are calculated in terms of equivalent quantities of calcium carbonate; a convenient method of reducing all salts to a common basis for comparison. Permanent hardeners remains in solution. Temporary hardness precipitates under favorable conditions to form scale.

HEAD LOSS: Reduction in liquid pressure associated with passage of a solution through a bed of exchange material; a measure of resistance of a resin bed to the flow of liquid passing through it.

HYDRAULIC CLASSIFICATION: Rearrangement of resin particles in an ion exchange unit. As backwash water flows up through a resin bed, particles are placed in a mobile condition causing larger particles to settle and smaller particles to rise to the top.

HYDRAZINE: A strong reducing agent having the formula N_2NNH_2 in the form of a colorless hygroscopic liquid. Used as an oxygen scavenger. It is very toxic.

HYDROGEN CYCLE: Term denoting operation of a cation resin that is regenerated with dilute acid.

HYDROXYL: Term used to describe the anionic radical (OH^-) that is responsible for alkalinity of a solution.

INFLUENT: Solution that enters any process.

INHIBITOR: Any compound that inhibits chemical activity (scale or corrosion) by filming or chemical modification.

ION: An atom or molecule that becomes charged by losing or gaining electrons. Solids dissolved in water dissociate into their electronic constituents. Cations have a positive charge and anions have a negative charge.

ION EXCHANGE (SOLID PHASE): A reversible process interchanging ions between a solid and a liquid with no substantial structural changes of the solid.

ION EXCHANGER: A component that uses the principles of ion exchanger to remove dissolved solids from water; also called a demineralizer.

IONIZATION: Dissociation of molecules into charged particles, or ions.

INORGANICS: These include salts and gases that dissociate freely and form positive and negative ions in water. Some common salts are chlorides, carbonates and sulfates of sodium, calcium and magnesium. The most common dissolved gas is carbon dioxide, which forms carbonic acid in water.

JACKSON CANDLE TURBIDITY: An empirical measure of turbidity based on the measurement of the depth of a column of water that is just sufficient to extinguish the image of a burning standard candle observed vertically.

LANGELIER'S INDEX: An indicator of whether water will either tend to dissolve or tend to precipitate calcium carbonate. To calculate Langelier's Index, the actual pH value of the water and Langelier's saturation pH value (pH_s) are needed. Langelier's saturation pH value is determined by the relationship between calcium hardness, total alkalinity, total solids concentration, and temperature of the water. Langelier's Index is determined from the expression $pH - pH_s$.

LEAKAGE: The phenomenon that occurs when some influent ions are not adsorbed or exchanged and appear in the effluent when a solution is passed through an under-regenerated exchange resin bed.

LIME SOFTENING: Addition of lime to water to remove hardness impurities, such as calcium and magnesium, by creating a precipitate that is removed as sludge.

LOCALIZED ATTACK: Form of corrosion where only specific areas on a metal surface are affected.

MACRORETICULAR: A term used to describe resins that have a rigid polymer porous network and that have existing pore structure even after drying. The pores are larger than atomic distances and are not part of a gel structure.

MAGNETITE: A self-limiting type of corrosion product of iron that occurs under low oxygen conditions; the oxide layer that is formed prevents further corrosion.

MAKEUP: Water in a cooling tower or boiler used to replenish the system after blowdown, leakage, or other losses. The effect is that of an infinite series of two-bed ion exchange units.

MIXED BED ION EXCHANGER: Anion and cation resin beads mixed together in an ion exchanger.

MOLECULAR WEIGHT; The sum of the atomic weights of all the consitutent atoms in the molecule of a compound.

NEGATIVE CHARGE: Electric potential an atom acquires when it gains one or more electrons; characteristic of an anion.

NEPHELOMETRIC TURBIDITY: An empirical measure of turbidity based on a measurement of the light scattering characteristics (Tyndall effect) of the particulate matter in the water.

NEUTRALIZE: The counteraction of acidity with an alkali, or of alkalinity with an acid. The combination forms a neutral salt.

NOBLE GAS OR METAL: An element that does not combine easily with other elements.

NON-VOLATILE TREATMENT: Type of water treatment describing addition of chemicals, such as phosphates, to water in order to control impurities; phosphate chemistry.

ONCE-THROUGH SYSTEM: Condenser circulating water system where water is drawn from a source, used in the system, and returned to the source.

OPEN SYSTEM: Water system that is exposed to atmosphere.

ORGANIC CONTAMINANTS: These can be man-made substances, such as herbicides, pesticides and detergents, or naturally occurring by-products of vegetative decay and biodecomposition such as humic and fulvic acids. They are all compounds that contain carbon and, generally, hydrogen.

ORTHOPHOSPHATE: A form of phosphate that precipitates rather than sequesters hard water salts.

OSMOSIS: Denotes the passage of water through a membrane from a lesser concentration to a higher concentration. (See also Reverse Osmosis.)

OXIDATION: Combination of any chemical compound or element with additional oxygen. In the case of iron, rust is the product of oxidation.

PART PER BILLION (ppb): Measure of proportion by weight, equivalent to a unit weight of solute per billion unit weights of solution.
Note: A part per billion generally is considered equivalent to one millionth of a gram per liter, but this is not precise. A part per billion is also equivalent to a milligram of solute per one thousand kilograms of solution.

PART PER MILLION (ppm): Measure of proportion by weight, equivalent to a unit weight of solute per million unit weights of solution.

Note: A part per million generally is considered equivalent to a milligram per liter, but this is not precise. A part per million is also equivalent to a milligram of solute per kilogram of solution.

PARTICULATE MATTER: Matter, exclusive of gases, existing in the nonliquid state that is dispersed to give a heterogeneous mixture.

pH: A measure of the acidity or alkalinity of a solution on a scale from 1 (most acidic) to 14 (most alkaline), where a pH of 7 is neutral.

PHYSICAL STABILITY: The quality which an ion exchange resin must possess to resist changes that might be caused by attrition, high temperatures and other physical conditions.

pK: An expression of the extent of dissociation of an electrolyte; the negative logarithm of the ionization constant of a compound.

pOH: An expression of the alkalinity of a solution; the negative logarithm of the hydroxyl ion concentration.

PHOSPHATE CHEMISTRY: See Non-Volatile Treatment.

PITTING: Form of localized attack that causes pits to be formed in metal.

POLAR MOLECULE: Molecule that exhibits equal and opposite charges on opposite sides.

POLYMERIZATION: The union of a considerable number of simple molecules called monomers, to form a giant molecule, known as a polymer, having the same chemical composition.

POLYPHOSPHATE: A form of phosphate that sequesters rather than precipitates hard water salts.

POROSITY: An expression of space available in ion exchange resins to liquids for large organic molecules.

POSITIVE CHARGE: Electrical potential acquired by an atom that has lost one or more electrons; a characteristic of a cation.

PRECIPITATE: Solid material that can no longer be held in a solution due to change in pressure, temperature, or chemical condition. It then precipitates out.

PRECOAT FILTER: A filter with two media: a precoat medium, such as diatomaceous earth, and a base medium.

PRESSURE DIFFERENTIAL: Difference in pressure across a filter medium; used to measure the efficiency of a filter; also called delta P.

PRIMING: Applied to boilers when water in the boiler drum becomes unstable and intense level changes occur in different parts of the drum. A wave actually travels back and forth in a boiler drum.

PURITY: Measure of the condition of water with respect to dissolved solids and gases.

QUALITY: Applied to steam to denote the amount of liquid carried over with the steam into the system. One hundred percent quality equals zero carryover.

QUATERNARY AMMONIUM: A specific basic group $[-N(CH_3)_4+]$ on which depends the exchange activity of certain anion exchange resins.

RADICAL: (Free Radical): An atom or compound in which there is an unpaired electron, as $H^{\cdot}$ or $^{\cdot}CH_3$.

RAW WATER: Untreated water.

RECIRCULATING SYSTEM: A system that continuously recirculates water.

REGENERANT: Solution used to restore activity in an ion exchanger. Acids are employed to restore a cation exchanger to its hydrogen form; brine solutions may be used to convert the cation exchanger to the sodium form. An anion exchanger may be rejuvenated by treatment with an alkaline solution.

REGENERATION: Restoration of activity in an ion exchanger by replacing ions adsorbed from treated solution by ions that were adsorbed on the resin.

RESIN BEAD: A small, porous (usually plastic) bead that contains electrically charged areas where ions are exchanged; may be either a cation bead or an anion bead, and may be either gel type or macroporous.

RESISTIVITY: A measure of resistance offered by a liquid to the flow of electrical current. Pure water essentially has infinite resistivity.

REVERSE DEIONIZATION: The use of an anion exchange unit and a cation exchange unit—in that order—to remove ions from solution.

REVERSE OSMOSIS: A process that applies pressure on a higher concentration side to cause water to pass through a membrane to a lower concentration side; i.e., the impurities are left behind. (See Osmosis.)

RINSE: Operation following regeneration; a flushing out of excess regenerant solution.

RYZNAR STABILITY INDEX: An indicator for predicting scaling tendencies of water, based on a study of operating results with water of various saturation indices.

SALT: An ionically bonded compound that is composed of a metal and a nonmetal.

SALT SPLITTING: Conversion of salts to their corresponding acids.

SATURATED SOLUTION: A solution that holds the maximum amount of solute.

SCALE: Deposit formed by precipitation of solids from solution onto a confining surface.

SEDIMENT: Any loose settled material in water or other liquid.

SELECTIVE CORROSION: A type of corrosion where a specific metal is corroded; other metals are not affected.

SEQUESTER: To isolate a metallic ion without having an actual chemical reaction. See Chelate.

SILICEOUS GEL ZEOLITE: A synthetic, inorganic exchanger produced by the aqueous reaction of alkali with aluminum salts.

SINGLE DISPLACEMENT REACTION: A chemical reaction in which an element replaces one of the elements in a compound.

SLUDGE: A relatively soft deposit, with a consistency similar to mud, that is formed from impurities.

SLUG: A fixed quantity of chemicals supplied to a system without time control.

SODIUM CYCLE: Term denoting operation of a cation resin when regenerated with salt.

SOLUTE: A dissolved substance in a solution.

SOLUTION: One substance dissolved in another; usually refers to solids in liquids, but may be liquid-liquid.

SOLVENT: Substance in a solution that causes another substance to dissolve.

STATIC SYSTEM: Batch use of ion exchange resins, where (since ion exchange is an equilibrium reaction) a definite endpoint is reached involving fixed ratios of ion distribution between the resin and solution.

SULFONIC: A specific acidic group controlling the activity of certain exchange resins.

SURFACE TENSION: A property arising from molecular forces of the surface film of all liquids which tends to alter the contained volume of liquid into a form of minimum superficial area.

SUSPENDED SOLID: An impurity that will not dissolve.

SWELLING: Expansion of an ion exchange bed that occurs when reactive groups on the resin are converted into certain forms.

THROUGHPUT VOLUME: Amount of solution passed through an exchange bed before exhaustion of the resin is reached.

TOTAL CHLORINE RESIDUAL: Total amount of chlorine residual present, without regard to type.

TOTAL MATTER: The sum of the particulate matter and dissolved matter.

TRIMER: A molecule derived from three identical simpler molecules, or a polymer derived from three identical monomers.

TURBIDITY: Reduction of transparency of water caused by particulate matter.

ULTRAFILTRATION: Separation of colloidal or very fine solid materials by filtration through membranes with pore sizes in the range of 10 angstroms to 1000 angstroms.

UPFLOW: Operation of an ion exchange unit where solutions are passed in at the bottom and out at the top.

VOIDS: Space between the resinous particles in an ion exchange bed.

VOLATILE MATTER: That matter that is changed under conditions of the test from a solid or a liquid state to the gaseous state.

VOLATILE TREATMENT: Type of water treatment involving addition of volatile chemicals to water to control impurities.

WATER-FORMED DEPOSITS: Any accumulation of insoluble material either derived from water or formed by the reaction of water on surfaces in contact with water.

ZEOLITE: Naturally occurring hydrous silicates exhibiting limited base exchange. In the past, used extensively for softening water.

Index